QUALITY ASSURANCE IN ANALYTICAL CHEMISTRY

Analytical Techniques in the Sciences (AnTS)

Series Editor: David J. Ando, Consultant, Dartford, Kent, UK

A series of open learning/distance learning books which covers all of the major analytical techniques and their application in the most important areas of physical, life and materials sciences.

Titles Available in the Series

Analytical Instrumentation: Performance Characteristics and Quality
Graham Currell, University of the West of England, Bristol, UK

Fundamentals of Electroanalytical Chemistry
Paul M.S. Monk, Manchester Metropolitan University, Manchester, UK

Introduction to Environmental Analysis
Roger N. Reeve, University of Sunderland, UK

Polymer Analysis
Barbara H. Stuart, University of Technology, Sydney, Australia

Chemical Sensors and Biosensors
Brian R. Eggins, University of Ulster at Jordanstown, Northern Ireland, UK

Methods for Environmental Trace Analysis
John R. Dean, Northumbria University, Newcastle, UK

Liquid Chromatography–Mass Spectrometry: An Introduction
Robert E. Ardrey, University of Huddersfield, Huddersfield, UK

Analysis of Controlled Substances
Michael D. Cole, Anglia Polytechnic University, Cambridge, UK

Infrared Spectroscopy: Fundamentals and Applications
Barbara H. Stuart, University of Technology, Sydney, Australia

Practical Inductively Coupled Plasma Spectroscopy
John R. Dean, Northumbria University, Newcastle, UK

Bioavailability, Bioaccessibility and Mobility of Environmental Contaminants
John R. Dean, Northumbria University, Newcastle, UK

Quality Assurance in Analytical Chemistry
Elizabeth Prichard and Vicki Barwick, LGC, Teddington, UK

Forthcoming Titles

The Analysis of DNA
Helen Hooper and Ruth Valentine, Northumbria University, Newcastle, UK

Techniques of Modern Organic Mass Spectrometry
Robert E. Ardrey, University of Huddersfield, Huddersfield, UK

Analytical Profiling and Comparison of Drugs of Abuse
Michael D. Cole, Anglia Polytechnic University, Cambridge, UK

QUALITY ASSURANCE IN ANALYTICAL CHEMISTRY

Elizabeth Prichard and Vicki Barwick
LGC, Teddington, UK

John Wiley & Sons, Ltd

Email (for orders and customer service enquiries): cs-books@wiley.co.uk
Visit our Home Page on www.wileyeurope.com or www.wiley.com

Reprinted with corrections April 2008

Other Wiley Editorial Offices

John Wiley & Sons Inc., 111 River Street, Hoboken, NJ 07030, USA

Jossey-Bass, 989 Market Street, San Francisco, CA 94103-1741, USA

Wiley-VCH Verlag GmbH, Boschstr. 12, D-69469 Weinheim, Germany

John Wiley & Sons Australia Ltd, 42 McDougall Street, Milton, Queensland 4064, Australia

John Wiley & Sons (Asia) Pte Ltd, 2 Clementi Loop #02-01, Jin Xing Distripark, Singapore 129809

John Wiley & Sons Canada Ltd, 6045 Freemont Blvd, Mississauga, Ontario, L5R 4J3, Canada

Wiley also publishes its books in a variety of electronic formats. Some content that appears in print may not be available in electronic books.

Anniversary Logo Design: Richard J. Pacifico

Cover image supplied by Charanjit Bhui, Senior Graphic Designer at LGC

Library of Congress Cataloging-in-Publication Data:

Prichard, F. Elizabeth (Florence Elizabeth)
Quality assurance in analytical chemistry / Elizabeth Prichard and Vicki Barwick.
p. cm.
ISBN 978-0-470-01203-1 (cloth)
1. Chemical laboratories – Quality control. 2. Chemistry, Analytic – Quality control. 3. Chemistry, Analytic – Technique. I. Barwick, Vicki. II. Title.
QD75.4.Q34P75 2007
543.0685 – dc22

2007011033

British Library Cataloguing in Publication Data

A catalogue record for this book is available from the British Library

ISBN 978-0-470-01203-1 (Cloth) 978-0-470-01204-8 (Paper)

Typeset in 10/12pt Times by Laserwords Private Limited, Chennai, India
Printed and bound in Great Britain by TJ International, Padstow, Cornwall
This book is printed on acid-free paper.

Contents

Series Preface

There has been a rapid expansion in the provision of further education in recent years, which has brought with it the need to provide more flexible methods of teaching in order to satisfy the requirements of an increasingly more diverse type of student. In this respect, the *open learning* approach has proved to be a valuable and effective teaching method, in particular for those students who for a variety of reasons cannot pursue full-time traditional courses. As a result, John Wiley & Sons, Ltd first published the Analytical Chemistry by Open Learning (ACOL) series of textbooks in the late 1980s. This series, which covers all of the major analytical techniques, rapidly established itself as a valuable teaching resource, providing a convenient and flexible means of studying for those people who, on account of their individual circumstances, were not able to take advantage of more conventional methods of education in this particular subject area.

Following upon the success of the ACOL series, which by its very name is predominately concerned with Analytical Chemistry, the *Analytical Techniques in the Sciences* (AnTS) series of open learning texts has been introduced with the aim of providing a broader coverage of the many areas of science in which analytical techniques and methods are increasingly applied. With this in mind, the AnTS series of texts seeks to provide a range of books which covers not only the actual techniques themselves, but *also* those scientific disciplines which have a necessary requirement for analytical characterization methods.

Analytical instrumentation continues to increase in sophistication, and as a consequence, the range of materials that can now be almost routinely analysed has increased accordingly. Books in this series which are concerned with the *techniques* themselves reflect such advances in analytical instrumentation, while at the same time providing full and detailed discussions of the fundamental concepts and theories of the particular analytical method being considered. Such books cover a variety of techniques, including general instrumental analysis, spectroscopy, chromatography, electrophoresis, tandem techniques, electroanalytical methods, X-ray analysis and other significant topics. In addition, books in this series also consider the *application* of analytical techniques in areas such as environmental science, the life sciences, clinical analysis, food science, forensic analysis, pharmaceutical science, conservation and archaeology, polymer science and general solid-state materials science.

Written by experts in their own particular fields, the books are presented in an easy-to-read, user-friendly style, with each chapter including both learning objectives and summaries of the subject matter being covered. The progress of the reader can be assessed by the use of frequent self-assessment questions (SAQs) and discussion questions (DQs), along with their corresponding reinforcing or remedial responses, which appear regularly throughout the texts. The books are thus eminently suitable both for self-study applications and for forming the basis of industrial company in-house training schemes. Each text also contains a large amount of supplementary material, including bibliographies, lists of acronyms and abbreviations, and tables of SI Units and important physical constants, plus where appropriate, glossaries and references to literature and other sources.

It is therefore hoped that this present series of textbooks will prove to be a useful and valuable source of teaching material, both for individual students and for teachers of science courses.

Dave Ando
Dartford, UK

Preface

This book is a very much revised edition of the book, *Quality in the Analytical Chemistry Laboratory*, published as part of the ACOL Series about twelve years ago. This version covers in more detail the technical requirements of the International Standards adopted by many laboratories, i.e. method validation, traceability and measurement uncertainty. At the time of writing the ACOL book, analytical measurements were beginning to play an increasing role in many of the world economies. During the last ten years the number of countries wishing to trade internationally has increased. In addition, the need for reliable measurements, carried out in a cost-effective way, brings new demands on analytical scientists. Analysts are required to measure more quickly samples in many different matrices, often with the analyte present at a very low concentration. In addition, the customers of analytical laboratories are also requiring some third-party evidence that they can trust the work of the laboratory. This is why the sections dealing with International Standards and Proficiency Testing schemes have been extended. Increasingly, a knowledge of quality assurance and quality control is important. The principles of reliable measurements need to be integrated into a scientist's work early on in his/her career.

LGC has a strong interest in the quality of analytical measurements and played a lead role in the development of the UK's initiative on Valid Analytical Measurement (VAM). This work is supported by UK government as part of the National Measurement System (NMS) Chemical and Biological Metrology Programme.

The aims of the programme are as follows:

- to improve the quality of analytical measurements made in the UK;
- to facilitate mutual recognition of analytical data across international boundaries;

- to develop a robust and transparent infrastructure aimed at achieving international comparability and traceability of chemical and biochemical measurements.

These aims reach much further than the UK as evidenced by other international organizations which have developed, partly as a result of the work in VAM. They include Eurachem (a network of organizations in Europe having the objective of establishing a system for the international traceability of chemical measurements and the promotion of good quality practices) and CITAC (Co-operation on International Traceability in Analytical Chemistry). A wide variety of issues have been identified, all of which need to be addressed to ensure that analytical measurements made in different countries or at different times are comparable.

This book sets out to describe how 'best practice' can be achieved through adopting the VAM principles (see Chapter 2). It describes the international standards that can provide laboratories with a structure to work to, as well as third-party assurance of their competence. It takes the analyst through the complete analytical process. The text starts with defining the problem fully, thus enabling analysts to identify a method that is likely to provide results that are 'fit for purpose'. This is ensured through appropriate method validation and suitable calibration leading to results that are traceable to international standards. This is followed by sampling and sample pre-treatment. Layout of the laboratory and inappropriate choice of chemicals can result in poor data. The internal quality control and external quality assessment are both described, as well as sections covering evaluation of the uncertainty of the measurement results. The analytical results end with the report being written and this is described with all of the other documents which ensure the smooth operation of a laboratory.

Issues of quality extend far beyond organizational arrangements. The analyst and the employer must both accept the equally important aspect of quality in professional skills and competences. Analysts need more help in order to learn about good laboratory practice and to work competently and professionally on a day-to-day basis within the framework of the VAM principles. This book offers analysts a new learning route to achieving these aims, and employers a convenient way to introduce quality assurance procedures.

The production of this book has been made possible through support from the VAM programme. The technical content of the book has also benefited through the results of work which has been carried out by LGC, often in collaboration with others, on specific technically based VAM, Eurachem and CITAC projects.

Elizabeth Prichard and Vicki Barwick

Acknowledgements

The authors gratefully acknowledge the assistance of the following, without whom this book would not have been completed. Reviews of specific sections were carried out by LGC staff, including Charlotte Bailey, Nick Boley, Celine Wolff-Briche, Brian Brookman, John Day, Alan Handley, George Merson and Graham Savage. In addition, the whole book was thoroughly reviewed by Dr Piotr Robouch, IRMM, Geel, Belgium and by Professor Alan Townshend, University of Hull. We are grateful for their hard work and valuable suggestions.

The authors take full responsibility for any errors or omissions.

The preparation of this book was supported under contract with the Department of Trade and Industry (DTI) as part of the National Measurement System (NMS) Valid Analytical Measurement (VAM) Programme. The work of the VAM Programme now forms part of the NMS Chemical and Biological Metrology Programme which is funded by the Department for Innovation, Universities and Skills (DIUS). The VAM Programme aims to improve the quality of analytical measurement in the UK.

Acknowledgements

Acronyms, Abbreviations and Symbols

°C	temperature on Celsius scale
abv	alcohol by volume
Ac	acceptance number in a sampling plan
ANOVA	analysis of variance
AOAC	Association of Official Analytical Chemists
AQL	acceptance quality limit
ASTM	American Society for Testing and Materials
B	measurement bias
BAM	Federal Institute for Materials Research and Testing (Germany)
BCR	Bureau Communautaire de Référence
BIPM	Bureau International des Poids et Mesures (Bureau of Weights and Measures)
BP	British Pharmacopeia
BS	British Standard
BSI	British Standards Institute
CEN	European Committee for Standardization
CFR	Code of Federal Regulations
CI	confidence interval
CITAC	Co-operation on International Traceability in Analytical Chemistry
CONTEST	Contaminated Land Proficiency Testing Scheme
COSHH	Control of Substances Hazardous to Health
CRM	Certified Reference Material
CUSUM	cumulative sum
DIUS	Department for Innovation, Universities and Skills

DTI	Department of Trade and Industry
EC	European Commission
EFQM EEA	European Foundation for Quality Management Excellence Award
EP	European Pharmacopeia
EPTIS	A Proficiency Testing information system
EQA	external quality assessment
ERM	European Reference Material
Eurachem	European network of analytical chemistry laboratories
FAPAS	Food Analysis Performance Assessment Scheme
FDA	Food and Drug Administration (USA)
GC	gas chromatography
GC–ECD	gas chromatography with electron-capture detection
GC–FID	gas chromatography with flame-ionization detection
GLP	Good Laboratory Practice
GNP	gross National product
HMSO	Her Majesty's Stationery Office
HPLC	high performance liquid chromatography
HPTLC	high performance thin layer chromatography
ICP–MS	inductively coupled plasma–mass spectrometry
ICP–OES	inductively coupled plasma–optical emission spectroscopy
IDL	instrument detection limit
IEC	International Electrotechnical Commission
ILAC	International Laboratory Accreditation Cooperation
IMEP	International Measurement Evaluation Programme
IP	Institute of Petroleum (now called the Institute of Energy)
IQC	internal quality control
IRMM	European Commission, Joint Research Centre, Institute for Reference Materials and Measurements (Belgium)
ISO	International Organization for Standardization
IUPAC	International Union of Pure and Applied Chemistry
LC	liquid chromatography
LEAP	Laboratory Environmental Analysis Proficiency
LIMS	Laboratory Information Management System
LoD	limit of detection
LoQ	limit of quantitation
MAD	median absolute deviation
MAD_E	estimate of robust standard deviation
MAPS	Malt Analytes Proficiency Testing Scheme
MDL	method detection limit
MS	mass spectrometry
MS^n	multiple stages of mass spectrometry
MU	measurement uncertainty

NIST	National Institute of Standards and Technology
NPD	nitrogen–phosphorus detector (also known as a thermionic detector)
OECD	Organization for Economic Co-operation and Development
PAH(s)	polyaromatic hydrocarbon(s)
PT	Proficiency Testing
QA	Quality Assurance
QC	Quality Control
QP	Quality Procedure
QMS	Quality Management's Quality in Microbiology Scheme
QWAS	Quality Management's Quality in Water Analysis Scheme
R	recovery ratio
Re	rejection number in a sampling plan
RICE	Regular Interlaboratory Counting Scheme
RM	reference material
SI	Système International (d'Unitès) (International System of Units)
SIM	selected-ion monitoring
SOP	Standard Operating Procedure
SRM	Standard Reference Material
TLC	thin layer chromatography
UKAS	United Kingdom Accreditation Service
USP	United States Pharmacopeia
UV	ultraviolet
VAM	Valid Analytical Measurement (Programme)
VIM	International Vocabulary of Metrology – Basic and General Concepts and Associated Terms (ISO/IEC Guide 99)
VIS	visible
WASP	Workplace Analysis Scheme for Proficiency
WI	Work Instruction

CV	coefficient of variation
E_n	score in PT scheme
k	coverage factor used to calculate the expanded uncertainty
K_s	sampling constant
n	number in a sample data set
N	number in a population of results
r	repeatability and correlation (regression) coefficient
R	reproducibility
RSD	relative standard deviation
RSZ	rescaled sum of z-scores
s	sample standard deviation
s^2	sample variance
s_r	repeatability standard deviation

s_R	reproducibility standard deviation
$s(\overline{x})$	standard deviation of the mean
SDM	standard deviation of the mean
SSZ	sum of squared z-scores
t	Student t-value
T	target value
u	standard uncertainty
$u(a)$	standard uncertainty of a
u_x	standard uncertainty of x
U	expanded uncertainty
$\overline{x}$	sample mean
x_i	datum point i
x_0	accepted reference value
X	assigned value in PT scheme
$\overline{y}$	mean response (of a calibration set)
z	score in PT scheme
z'	score in PT scheme
α	probability of a value being outside of a specified range
θ	angle used in construction of CUSUM V-mask
μ	population mean
ν	number of degrees of freedom
σ	population standard deviation
σ^2	population variance
σ'	between-laboratory standard deviation
σ_r	repeatability standard deviation
σ_R	reproducibility standard deviation
$\hat{\sigma}$	target range in PT scheme

About the Authors

Elizabeth Prichard, B.Sc., Ph.D., FRSC, CChem.

Elizabeth Prichard obtained a first degree in Chemistry from the University College of Wales, Aberystwyth, where she went on to obtain her doctorate studying infrared spectroscopy. After a Civil Service Research Fellowship, she moved into academia, initially at Bedford College and then Royal Holloway and Bedford New College, University of London before moving to the University of Warwick as a Senior Research Fellow. While at London University, she continued her research in spectroscopy, as well as some work in biophysical chemistry. At the University of Warwick, she researched the release profiles of steroids from implanted contraceptive devices. During her time at London University, Elizabeth spent sabbatical periods at the Division of Materials Application, NPL, the Biophysics and Biochemistry Department, Wellcome Research Foundation, Beckenham and then as Associate Professor sponsored by the British Council at the University of Gezira and at the University of Khartoum, Sudan.

In 1992, Elizabeth was seconded from the University of Warwick to the Laboratory of the Government Chemist (which became LGC in 1996) to manage the production of the first edition of this book which was part of the ACOL Series. Elizabeth had been an editor for this series since its inception in the late 1980s. In 1997, Elizabeth joined the staff of LGC as Head of Education and Training. She developed there a Quality Assurance Training Programme and continues to deliver these courses. As part of the Valid Analytical Measurement (VAM) programme, she also developed courses in analytical science appropriate for A-level chemistry teachers and a Proficiency Testing (PT) competition for the students. She has been involved in the production of a number of quality assurance products for all levels of learning from school to the professional analyst, including books, videos, CDs and web-based e-learning material. In addition, Elizabeth has been a partner in EU projects in the area of quality assurance – these include the QUACHA and SWIFT-WFD projects.

During the last fifteen years, Elizabeth has lectured and delivered training courses on quality assurance topics in the UK, Europe and USA.

Vicki Barwick, B.Sc

Vicki Barwick obtained a first degree in Chemistry from the University of Nottingham. She then joined the Laboratory of the Government Chemist (which became LGC in 1996) as an analyst in the Consumer Safety Group. Vicki was involved with a number of projects to assess the safety of consumer products, including developing test methods for the identification of colourants in cosmetics and the quantitation of phthalate plasticizers in child-care items.

After five years as an analyst, Vicki moved within LGC to work on the DTI-funded Valid Analytical Measurement (VAM) programme. In this role, she was responsible for providing advice and developing guidance on method validation, measurement uncertainty and statistics. One of her key projects involved the development of approaches for evaluating the uncertainty in results obtained from chemical test methods. During this time, Vicki also became involved with the development and delivery of training courses on topics such as method validation, measurement uncertainty, quality systems and statistics for analytical chemists.

In 2002, Vicki moved into a full-time education and training role. She is currently the project manager for knowledge transfer projects under the National Measurement System Chemical and Biological Metrology Programme. These projects aim to promote the adoption of the principles of valid analytical measurement. This involves the production of training resources and the organization of workshops for analysts. Vicki is also responsible for the development and delivery of LGC's Quality Assurance Training Courses.

Over the last ten years, Vicki has lectured extensively on quality assurance topics and has co-authored a number of papers, books and guides.

Chapter 1
The Need for Reliable Results

Learning Objectives

- To understand why analytical measurements need to be made.
- To understand the importance of producing reliable results.
- To be able to define what is meant by 'quality'.

1.1 Why Analytical Work is Required

Measurements affect the daily lives of every citizen. Sound, accurate and reliable measurements, be they physical, chemical or biological, are essential to the functioning of modern society. For these reasons, advanced nations spend up to 6% of their Gross National Product (GNP) on measurements and measurement-related operations.

Informed debate and decisions on such important matters as the depletion of the ozone layer, acid rain and the quality of waterways all depend on the data provided by analytical chemists. Forensic evidence also often depends on chemical measurements. National and international trade are critically dependent on analytical results. Chemical composition is often the basis for the definition of the nature of goods and tariff classification. In all of these areas not only is it important to get the right answer but it is essential that the user of the results is confident and assured that the data are truly representative of the sample and that the results are defendable, traceable and mutually acceptable by all laboratories.

The role of the analytical chemist has not changed since the time analysts discovered that naturally occurring products were composite materials. For example,

Quality Assurance in Analytical Chemistry E. Prichard and V. Barwick

when it was discovered that carrots helped prevent 'night blindness' it was an analytical chemist who separated out the various components of carrot, characterized the compounds and identified the active component as ß-carotene. What have changed are the questions that are asked by society; they have become more demanding. Much of the interest today is centred on levels of unwanted materials that are present at a very low concentration, such as ng g^{-1}. In addition, the range of materials being analysed has increased enormously, with the results being required very quickly, as cheaply as possible and to the best quality.

1.2 Social and Economic Impact of a 'Wrong Analysis'

The social and economic impact of the analyst getting a wrong result and the customer consequently reaching a false conclusion can be enormous, for example:

- In forensic analysis, it could lead to a wrongful conviction or the guilty going unpunished.
- In trade, it could lead to the supply of sub-standard goods and the high cost of replacement with subsequent loss of customers.
- In environmental monitoring, mistakes could lead to hazards being undetected or to the identification of unreal hazards.
- In the supply of drinking water, it could lead to harmful contaminants being undetected.
- In healthcare, the incorrect medication or the incorrect content of active ingredient in a tablet can be catastrophic for the patient.

Just think of the huge costs, both in terms of financial and other resources, and in terms of the distress to individuals and their families, that could be caused by such mistakes. In all areas of application 'getting it wrong' leads to loss of confidence in the validity of future analytical results. Confidence is an important commodity. At one extreme, loss of confidence puts the future existence of the particular analytical laboratory at risk, but more generally it leads to costly repetition of analyses and, in the area of trade, inhibits the expansion of the world economy.

Many of you will be able to call to mind reports in the papers, or on TV, where the analytical chemist has apparently made a mistake. Some of these may be notorious but remember the many million times that the analytical chemist gets it right without any publicity. We are all aware of the debate over global warming; just think of how important it is that action taken in the future is based on information that gives a true picture of the composition of the global atmosphere. This book, by covering the majority of relevant basic issues, is designed to put you on the right path to quality in the analytical laboratory.

You might think that results obtained these days are more reliable than they were in the past. This may be true. The technologies have improved, tools for quality control, e.g. Certified Reference Materials, are available and new, more specific quantitative methods have been developed. However, there is ample proof that there are data being produced that are not fit for their intended purpose. Much of the evidence for unreliable results has come from studies involving a number of experienced laboratories all measuring samples of the same material. These may be from studies called collaborative studies or from the results produced during rounds of a Proficiency Testing Scheme (see Chapter 7) or from results published by the International Measurement Evaluation Programme (IMEP) [1]. This programme was set up to demonstrate the degree of equivalence of results of chemical measurements on a global scale. Figure 1.1 shows results for cadmium analysis from IMEP 9 where the participating laboratories have analysed samples of water containing trace metals. It is clear from this figure that many laboratories are not producing results which are 'fit for purpose'. This can be because of human error or because the results are not linked to a traceable reference value.

The issue is that the level of quality control that analysts have applied to their measurements in the past has been insufficient to meet the new challenges of today's analytical problems. There are many reasons why a laboratory might

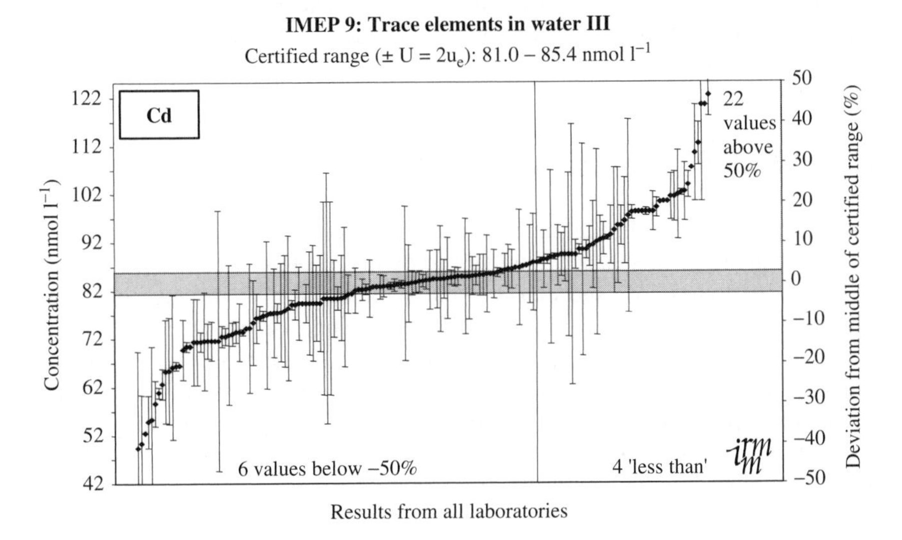

Figure 1.1 Results from an international intercomparison, IMEP 9, where U is the expanded uncertainty and u_e the combined standard uncertainty [1] (see Chapter 6, Section 6.3). Reproduced by permission of EC-JRC-IRMM (Philip Taylor) from IMEP 9, 'Trace elements in water III, Cd, certified range 81.0–85.4 nmol l^{-1}'.

quote incorrect results. A result may be incorrect because of an error in a calculation or an instrument that is out of calibration so that the scale reading no longer shows the correct value. However, more often the error is in the method used for the analysis. If the method used is not suitable for the analysis then the result will be incorrect. This could be because the analyte concentration is outside the range for which the method has been validated (see Chapter 4, Section 4.6). The method may be unable to detect the low levels required. Another reason could be that there is an interfering substance present, which is being detected along with the analyte. This is termed, 'lack of selectivity'. A more subtle reason may be that the method does not measure exactly what the customer requires. The method may measure 'total iron' in a tablet although the customer wished to know the amount of iron that is extracted by stomach acid. In addition, the method used may be very sensitive to some small changes that have been made to a validated method, e.g. in the concentration or the amount of material added, scaling up/down of components, temperature or pressure, etc. The extent to which a method can be modified without significant loss of accuracy is a measure of the 'ruggedness' of the method (see Chapter 4, Section 4.6.5).

Whatever the reason for obtaining it, a chemical measurement has a certain importance since decisions based upon it will very often need to be made. These decisions may well have implications for the health or livelihood of millions of people. In addition, with the increasing liberalization of world trade, there is pressure to eliminate the replication of effort in testing products moving across national frontiers. This means that quantitative analytical results should be acceptable to all potential users, whether they be inside or outside the organization or country generating them.

1.3 What do we Mean by 'Quality'

This book is about quality in the analytical chemistry laboratory, but what do we mean by 'quality'? It is easier to understand when dealing with various products, e.g. cars or clothes. All successful manufacturers have to produce goods that they can sell. Car manufacturers will have a range of products to suit their customers' needs. They will all be made to a high standard so that they comply with legislation; however, they will be aimed at people with different needs. You can compare this with an analytical laboratory. Analytical chemists produce results that are passed on to someone else (the customer) who will use them to solve a problem. The laboratory is providing a service.

DQ 1.1

What do you think quality means in terms of the results obtained in the analytical laboratory?

Answer

Quality in this context is not necessarily getting the most accurate results – it is matching the service with the requirements of the customer. This is achieved by providing results that:

- meet the specific needs of the customer;
- attract the confidence of the customer and all others who make use of the results;
- represent value for money.

This is often referred to as *fitness for purpose*.

DQ 1.2

List the factors you think need to be considered to ensure that you provide a customer with results of acceptable quality.

Answer

There are a number of things to consider, but the most important is understanding the needs of the customer. Is the total sugar content of the product required or the lactose content? The level of uncertainty in the result that is acceptable also helps focus on the choice of method. Once the method is chosen and validated, it is then important to ensure that all of the equipment is available and in a proper state of calibration. Then, all that remains is to have sufficient trained staff to carry out the analysis. Once the experimental results have been obtained and the data treatment is complete, the report can be written. The report also has to meet the customer requirements and should be written in an unambiguous way which is clear to the non-specialist.

1.4 Customer Requirements

To ensure that analytical results are fit for purpose, there has to be a discussion with the customer before the analysis is started. You must remember that a customer who is a member of your laboratory is just as important as the customer from outside your organization.

It goes without saying that you should make all measurements to the best of your ability. However, a value to the highest level of precision and trueness is not always required. The aim is that the result produced should be accurate enough to be of use to the customer, for the intended purpose (see Chapter 4). Customers may want the technical details of the method used but more often this will not be required. It is therefore vital that the exact requirements are discussed with the customer prior to the analysis and mutually agreed. The customer will require enough evidence to give confidence that the data are accurate and are suitable for their intended purpose. The data need to be backed-up by documentary evidence, such as computer printouts and record books that have been signed and dated, and checked. These documents may be required as evidence in cases of disputes or complaints. Every result you produce should have included with it an estimate of the uncertainty associated with the value (see Chapter 6).

There are several different categories of analysis, which might be required by a customer. Each requires a different approach. Analysing to a 'specification' where there is a maximum or minimum limit, for which a product or component concentration passes or fails, requires a different analytical approach from that required for a screening method requiring a 'yes/no' analysis. However, the customer requirement of fitness for purpose stays the same. A screening or 'yes/no' method is used when you have a large number of samples so you need a quick method to select which ones should be subjected to additional testing. The guidance on the maximum level of arsenic in contaminated land is 40 mg kg^{-1}. The analysis needs to be quantitative, accurate and reproducible at the 40 mg kg^{-1} level. However, the method does not need to be accurate over a wide range of concentrations of the analyte to be determined, e.g. 1 mg kg^{-1} to 100 mg kg^{-1}. If the land is contaminated above the guidance limit, it does not matter whether it is 45 mg kg^{-1} or 145 mg kg^{-1}; the land is 'condemned'. Similarly, if the concentration is less than, say 10 mg kg^{-1} it does not matter if the error is 100%. Where the customer does need assurance is how reliable the information is and what confidence can be placed on the data at or around the 40 mg kg^{-1} level. Is 41 mg kg^{-1} or 39 mg kg^{-1} unacceptable or acceptable? All of the procedures have to be fully documented (see Chapter 8), so that answers can be found to the following questions:

- What is the precision and accuracy of the method?
- Was the method tested with known samples to show that it was suitable in terms of the analyte and the concentration range, i.e. validated?
- What data are available concerning sampling, extraction procedures and the end measurement?

Forensic analysis is usually required for the collection of data in the course of determining whether legislation has been infringed. The customer requires that, above all, there is an unbroken chain of evidence from the time the samples were taken to the presentation of evidence in courts of law. In the laboratory this will include documentation and authorization for sample receipt, sample transfer, sub-sampling, laboratory notebooks, analytical procedures, calculations and observations, witness statements and sample disposal. All of these aspects can be called as evidence in court.

Every analytical chemist should be asking the same questions, i.e. 'Am I using a method which is appropriate and fit for purpose, has it been validated and what are the sources of uncertainty in the method and in my technique? What confidence do I have in the final answer?' Scientists are increasingly replacing the notion of *error* by that of *uncertainty*. Just because an instrument shows a scale reading of 3.4276 does not mean that the figures are all true and are known to the same level of certainty. It will depend on the state of calibration and on the machine having been used in a proper manner. Even the simplest titration will have a degree of uncertainty at several stages. For example, if a 25 ml burette is used in an acid–base titration, the reading on the burette may be 10.50 ml at the end-point. However, there is an associated uncertainty from two sources in this reading. First, there is an uncertainty in the visual measurement due to parallax and the interpretation of the meniscus and secondly the uncertainty of the calibration of the burette itself. This will depend on the grade of burette used. The reason for this and the way to go about making an assessment of uncertainty in a chemical measurement is explained in Chapter 6.

As an analyst you understand the meaning of the scientific data you produce. However, it must be remembered that laymen often do not and so the data need to be documented in a form that is easily understood. For example, the chromatographic analysis of hydrocarbon oil from an oil spill can produce a chromatogram with over 300 components. Explaining the significance of such data to a jury will be of little benefit. However, overlaying it with a standard trace can demonstrate pictorially whether there is a similarity or not. The customer requires information from the analyst to prove a point. If the data are not fully documented, then the point cannot be proven. A customer who has confidence in a laboratory will always return.

1.5 Purpose of Analysis

Analysis involves the determination of the composition of a material, i.e. the identification of its constituent parts and how much of each component is present and, sometimes, in what form. Before starting work on a sample, it is vital to enquire why the work is required, what will happen to the result(s) and to find out what decisions will be taken based on the numerical values obtained.

The purpose of an analysis and the use to which the analytical report or certificate of analysis might be put are numerous. A few examples are listed below:

- Preparation of a data bank of figures to establish trends, e.g. changes in pesticide residue concentration in foods with season, or from year to year.
- Acceptance/rejection of a chemical/product used in a manufacturing operation.
- Assessment of the value of a consignment of goods before payment.

- Prosecution of a company for selling a product not up to the stated specification, e.g. a sausage containing insufficient meat or containing pork instead of beef.
- Providing evidence in a case for or against an individual charged with being in possession of illegal drugs.

In all cases, there could be serious consequences depending on the particular investigation.

An error in the data bank figures may become apparent as further work is completed. If the error is a simple calculation error it can be corrected. However, if a mistake was made because of selection of an unsuitable method or in the calibration of instruments or in the choice of reagents used, it may not be possible to correct the error. This is particularly true if the original samples have been used up or if they have deteriorated in storage. Nevertheless, the error may not be serious where trends are under investigation, e.g. trends over a period of time or trends produced as a result of different treatments. This is because the absolute value of the measurement is of far less importance than the change from day to day, treatment to treatment, etc. Hence, so long as errors remain constant, differences between results will be real. This may not be true if different methods and/or different equipment are used or when trends are being monitored by a number of different laboratories.

Acceptance/rejection or valuation cases may cost (or save) a company a great deal of money depending on the error in the analysis and size of production runs. The prosecution of a company may give rise to a fine, or in the most severe cases, imprisonment of individuals. The arrest of an individual for possession of drugs (or explosives) could have very serious consequences; the individual concerned may be convicted. If the identification of the substance was made in error, the convicted person will have suffered unnecessarily and there could subsequently be huge compensation claims.

Hence, the choice of method and the validation of the method selected become increasingly more critical for those analyses resulting in actions towards the bottom of the list than for those at the top. You now need to pause to consider what the consequences of poor analytical work could be in your own particular job. Do not forget to include longer-term implications as well as the immediate problems. Poor or wrong data also result in the loss of reputation – customers who never come back.

Summary

This chapter sets the scene for the rest of the book. It gives examples of why analytical measurements are made. This then leads on to why the reliability of these results is so important. Quality is often a misunderstood word in the context of services and products. This chapter explains what it means in terms of analytical results.

Reference

1. IMEP 9, 'Trace Elements in Water III, Cd, Certified Range 81.0–85.4 nmol l^{-1}', European Commission, Joint Research Centre, Institute for Reference Materials and Measurements (EC-JRC-IRMM), Geel, Belgium (1998). [www.irmm.jrc.be/html/interlaboratory_comparisons/imep/index.htm] (accessed 30 November, 2006).

Chapter 2

General Principles of Quality Assurance and Quality Control

Learning Objectives

- To appreciate the need for quality assurance.
- To understand the importance of setting up a quality management system.
- To be able to define what is meant by quality assurance and quality control.
- To be aware of some of the international quality standards.

2.1 Introduction to Quality Assurance

The analyst provides scientific evidence on which important decisions are made. The work of an analyst is devalued if it is merely considered to be making measurements and reporting results. There has to be some added value. This is provided when it can be demonstrated that the results produced have been obtained in an organization that operates a quality management system. It is because of the importance of the work that the quality has to be assured. This means that all of the necessary actions have been taken to make sure that any factor which has an influence on the final result has been considered, understood and reported in a permanent record – that the appropriate measurements have been made and these have been carried out correctly using a validated method.

No one deliberately produces incorrect results. You may have noticed how often people remind you of the mistakes you have made but rarely of the good work. If your results prove to be wrong, it is not just your reputation that will

Quality Assurance in Analytical Chemistry E. Prichard and V. Barwick

Table 2.1 Trace metals in sea water ($\mu g\ l^{-1}$)

Metal	1965	1975	1983
Lead	0.03	0.03	0.002
Mercury	0.03	0.03	0.001
Nickel	2.0	1.7	0.46
Copper	3.0	0.5	0.25
Zinc	10.0	4.9	0.39

suffer, your customers and subsequently your organization will also suffer. Everyone makes mistakes but when you do, you must always try to find out why it happened, thereby reducing the chances of it happening again.

There is ample proof in the literature that there are data being produced that are not fit for their intended purpose. Table 2.1 shows the accepted representative values of the concentration of trace metals in the open ocean over a 20-year span. The values given might suggest that there has been a dramatic decrease in the level of these metals, e.g. the lead levels appear to have dropped 15-fold. The reasons for the apparent reduction in metal content in the sea water over the period studied could be a reduction in polluting materials, change in sea flow or improvement in the selectivity of the analytical technique. However, the concentration of metals in open ocean water is expected to remain fairly constant over time. There are a number of problems with this type of study. These include difficulty of sampling and storing of samples and the delays in carrying out the analysis, as well as the analysis itself. Unfortunately, there is no way of giving a definitive answer as to why laboratories have reported values which may differ by a factor of 10^4. There is insufficient documentary evidence of how the earlier measurements were made and one cannot get a 1965 sample to repeat the measurement! Therefore, there is no basis from which to draw conclusions.

It is important that a measurement made in one laboratory by a particular analyst can be repeated by other analysts in the same laboratory or in another laboratory, even where the other laboratory may be in a different country. We aim to ensure that measurements made in different laboratories are comparable. We are all confident that if we measure the length of a piece of wire, mass of a chemical or the time in any laboratory, we will get, very nearly, the same answer, no matter where we are. The reason for this is that there are international standards of length, mass and time. In order to obtain comparable results, the measuring devices need to be calibrated. For instance, balances are calibrated by using a standard mass, which can be traced to the primary mass standard (see also Chapter 5). The primary standard in chemistry is the amount of substance, i.e. the mole. It is not usually possible to trace all of our measurements back to the mole. We generally trace measurements to other SI units, e.g. mass as in 40 mg kg^{-1}

or trace back to reference materials which are themselves traceable to SI units. In analytical chemistry, we do not have a standard mole. Therefore, solutions made up to a well-defined concentration using very pure chemicals are used as a basis from which we can compare other solutions or an instrument scale. This process is 'calibration'. For some analyses, the chemical used may be a Certified Reference Material which has a well documented specification, e.g. in terms of the concentration of a particular species and the uncertainty of the specified value. However, it is not sufficient just to calibrate the apparatus/equipment used, it is important that the complete method of analysis is validated from extraction of the analyte from the sample to the final measurement.

The biggest benefit of producing reliable and traceable data is the mutual acceptance of those test data both nationally and internationally, by manufacturers, regulators, traders and governments. In this context, metrological traceability is as defined in the *International Vocabulary of Metrology* (VIM) [1]. Metrological traceability is the property of a measurement result whereby the result can be related to a reference through a documented unbroken chain of calibrations, each contributing to the measurement uncertainty. In terms of international trade, it is important that countries within a trading community accept each other's results. This can only happen if all of the countries agree the standard of quality testing and calibration. To achieve this, everyone must use the same reference points and be third-party assessed.

The increasing concerns of the public and the need for monitoring very low concentrations of toxic compounds means that detection at levels below $\mu g\ kg^{-1}$ are required in many areas of analysis. Pesticides in the food chain, toxic materials in incineration and waste products and traces of nitro-compounds in 'finger washings' of a person suspected of handling explosives, all involve analysis for low concentrations.

If data are produced which are not fit for their intended purpose, there is both the financial penalty and the possible legal penalty to be considered. For instance, if a manufacturer of pharmaceuticals produces a pharmaceutical tablet containing the incorrect amount of active ingredient, the consequences could be disastrous, causing, in the worst circumstances, loss of life.

Contaminated land is a 'never to be forgotten' high profile area of environmental pollution. However, to do anything with the land, so that it can be made fit for future uses, assessment has to be made on the basis of reliable analytical data. If the sampling and analysis are performed appropriately, then a set of results can be produced which can be used with confidence to make decisions about how the land can be made suitable for use. Random or unintelligent sampling can miss patches or 'hot spots' of pollution, which can have disastrous effects (see Chapter 3 for a discussion of sampling). If bad sampling has resulted in a high concentration of phenols or sulfate from an old gas works being undetected, the concrete in the piles for a new multi-story building could be attacked with the eventual weakening of the structure. It is also possible that if assessment of land is not performed reliably, then if the land is used for gardens, the vegetables grown on that land may provide a source of toxic metals which can end up in the food chain.

The benefits of producing good data are therefore broad and impinge on all of our daily lives, whether it is food, environment, health or trade. Laboratories that produce valid measurements have a higher status in the analytical world, since they produce data that are demonstrably traceable to a reference standard and reliable, with the cost of correcting bad data being lower. This means that such laboratories have a better chance of competing in the open market.

2.2 Quality Management System, Quality Assurance (QA) and Quality Control (QC)

Quality appears to have a language of its own and some definitions or explanations are required. This is what is covered in this section. A quality management system is a set of procedures and responsibilities that an organization puts in place to make sure that the staff have the facilities, equipment and resources to carry out their work effectively and efficiently. For the quality management system to be formally recognized and audited, it must be based on an internationally recognized standard that means something to its customers and other organizations across the world. There are a number of aspects of a good management system that should be put in place. These include the quality policy statement, general organization of the laboratory, roles and responsibilities, quality procedures, document control and reporting of results, auditing, review and subcontracting. It should be remembered that a quality management system should only be as comprehensive as that which is required to meet the needs of the customers. This means it may exceed the expectation of any individual customer.

Briefly, the quality management system is a combination of quality management, quality control and quality assurance. Quality assurance and quality control are components of the laboratory's quality management system. There is often confusion over the meaning of quality control and quality assurance and regrettably they are often used interchangeably. This is possibly because some quality control and quality assurance actions are interrelated. The definition of the terms can be found in the International Organization for Standardization (ISO) Standard, ISO 9000:2005 [2].

Quality assurance is the part of quality management focused on providing confidence that quality requirements will be fulfilled. It is all the planned and systematic activities implemented within the quality system, and demonstrated as needed, to provide adequate confidence that the analytical service will fulfil the requirements for quality. Quality assurance is the essential organizational infrastructure that supports all reliable analytical measurements. It encompasses a number of different activities. Included are staff training, record keeping, appropriate laboratory environment for the particular activities, adequate storage facilities to ensure the integrity of samples, reagents and solvents, maintenance and calibration schedules for instruments and the use of technically validated and documented methods.

Quality control is the part of the quality management system focused on fulfilling quality requirements. It is the planned activities designed to verify the quality of the measurement, e.g. analysing blanks or samples of known concentration. There are two types of quality control – internal quality control and external quality assessment (also known as external quality control). Internal quality control provides confidence to the laboratory management while external quality assessment provides confidence to the customer. Internal quality control is the operations carried out by the staff as part of the measurement process which provide evidence that the system is still operating satisfactorily and the results can be accepted (see also Chapter 5). Evidence is required that the performance of the laboratory is not only consistent but also comparable with other laboratories making similar measurements; this is provided by external quality assessment. Laboratories taking part in formal or informal intercomparisons achieve this. Formal intercomparisons are termed Proficiency Testing (PT) (see Chapter 7).

2.3 Different Standards and their Main Features

Standards dealing with the quality of a service have been developed by a number of national and international organizations. The requirements of an analytical laboratory depend on its size, the range of its activities and the type of analysis carried out. There are therefore a number of Standards that should be considered by an analytical laboratory. These Standards will be covered in more detail in Chapter 9.

DQ 2.1

Is there a difference between the processes of certification and accreditation?

Answer

There is a subtle difference in the meaning of these two terms and it is important to be clear about this difference. This is explained in the following paragraphs in terms of two Standards which you may encounter in your work.

The International Standard, ISO 9001:2000, Quality Management Systems – Requirements, is a general standard that applies to all types of organizations, regardless of size [3]. This Standard specifies the requirements for a quality management system where an organization, (a) needs to demonstrate its ability consistently to provide a service that meets customer and applicable regulatory requirements, and (b) aims to enhance customer satisfaction through the effective application of the system, including processes for continual improvement of the system and the assurance of conformity to customer and applicable regulatory requirements. If this meets the needs of the organization, it will seek certification by a third-party against this Standard. Certification is the procedure by which an external, independent auditing body (third-party) gives written assurance that a product, process or service conforms to specified requirements [2]. Independent third-party certification is offered by, e.g. BSI British Standards, the UK National Standards Body and, on an international scale, BSI Management Systems or Lloyds Register of Quality Assurance. This Standard does not check the competence of the organization but the control of the processes.

Another International Standard, ISO/IEC 17025:2005, General Requirements for the Competence of Testing and Calibration Laboratories, is far more specific [4]. This Standard applies to all laboratories performing tests and/or calibrations. It covers the competence of the laboratory and its staff. Laboratories seeking accreditation to this Standard will develop their management system for quality, administrative and technical operations in line with the clauses in this Standard. Accreditation is the formal procedure carried out by the relevant authority, which confers formal recognition that a laboratory is competent to carry out certain tasks [5]. This includes the management issues dealing with administration and those quality issues that are within certification. Accreditation is usually for a specific combination of analyte, matrix and method, although there are opportunities for flexible scope. Guidance on the implementation and management of flexible scopes of accreditation is produced by the International Laboratory Accreditation Cooperation (ILAC G18) [6] and many national accreditation bodies, e.g. UKAS LAB 39 [7]. The national accreditation body usually carries out the accreditation in that country. The United Kingdom Accreditation Service (UKAS) is the competent body in the UK.

Medical laboratories have some specific needs and these are incorporated in ISO 15189:2007, Medical Laboratories – Particular Requirements for Quality and Competence [8]. The requirements of both ISO 9001 and ISO/IEC 17025 are incorporated within this Standard. It is a customized version of ISO/IEC 17025 for medical laboratories. In the UK, UKAS have designated Clinical Pathology Accreditation (UK) Ltd as the authoritative body to accredit against this Standard.

Test facilities in the OECD (Organization for Economic Co-operation and Development) member countries that conduct regulatory studies must comply with the OECD Principles of Good Laboratory Practice (GLP), as set out in Council Decision C(97)186/Final. These are referred to as GLP Principles. GLP came into prominence in the late 1970s in response to some malpractice in research and development activities of pharmaceutical companies and contract laboratories used by them. The malpractice included some cases of fraud, but the majority was a result of poor management and organization of the studies carried out as part of the registration process for prospective pharmaceutical products. Problems were first highlighted in the United States and as a consequence the Food and Drug Administration (FDA) developed, in 1976, a set of principles for such studies which had to be adhered to before a regulatory authority could accept data from the studies. A harmonized set of regulations was required to ensure that international trade was not impeded. These were produced by the OECD and the initial set of principles was adopted in 1981. These were subsequently revised and published in 1998. The purpose of the GLP principles is to promote the development of quality test data. The principles set out a quality system dealing with the organizational process and the conditions under which non-clinical health and environmental safety studies are planned, performed, monitored, recorded, archived and reported. There needs to be sufficient information available for the study to be reconstructed at a future date. In the European Union and many other parts of the world, e.g. the USA and Japan, it is a regulatory requirement that studies undertaken to demonstrate the health or environmental safety of new chemical or biological substances shall be conducted in compliance with the principles of GLP. Each country has to have a monitoring authority that assesses the study to ensure it meets the requirements of the GLP principles. In some countries there may be more than one authority. In the United Kingdom, the GLP requirements are contained within The Good Laboratory Practice Regulations, Statutory Instrument 1999 No. 3106 with the amendments set out in Statutory Instrument 2004 No. 994 [9, 10]. These regulations require that any test facility which conducts, or intends to conduct, regulatory studies must be a member of the UK GLP Compliance Monitoring Programme. In these regulations, a 'regulatory study' is defined as a non-clinical experiment or set of experiments:

(a) in which an item is examined under laboratory conditions or in the environment in order to obtain data on its properties or its safety (or both) with respect to human health, animal health or the environment;

(b) the results of which are, or are intended, for submission to the appropriate regulatory authorities;

(c) for which compliance with the principles of Good Laboratory Practice is required in respect of that experiment or set of experiments by the appropriate regulatory authorities.

The Department of Health has responsibility for the monitoring in the UK through the GLP Monitoring Authority.

2.3.1 Common Features of ISO 9001, ISO/IEC 17025 and ISO 15189

The three Standards, ISO 9001, ISO/IEC 17025 and ISO 15189, have much in common but there are subtle differences. Figure 2.1 shows the common features. These will be covered briefly here and in more detail in Chapter 9. In matters relating to the management structure and responsibility for quality matters, the Quality Manager (however named) usually acts on behalf of the Chief Executive. The management structure is laid out in the Quality Manual which is the top-level document. It shows clearly how the organization is structured and the roles and responsibilities of the managers and their teams. The Quality Manual also contains all of the top-level documents relating to central functions and possibly some business areas. This will be supported by local documentation, e.g. Work Instructions and Standard Operating Procedures. It is important that all matters relating to a particular sample can be tracked from contract review, through the receipt of the sample to the delivery of the results to the customer. This means that all of the instrument results, validation and calibration data and quality control results must be uniquely identified as relating to the sample. Such records will have to be kept for the period of time set down in the Quality Manual or as agreed with the customer. Increasingly, organizations are expected to be able to demonstrate the competence of their staff. How this is achieved may also be laid down in the Quality Manual. In order to demonstrate to the senior management that the appropriate procedures are in place and being implemented, internal checks are carried out on a regular basis – these are called internal audits. These provide a means of ensuring correct implementation of the quality system. In addition, to gain either certification or accreditation, to a particular standard, there has to be third-party assessments by appropriate authorities. To ensure that all the requirements of the particular Standard and the needs of the business are satisfied, the quality management system is reviewed by top management once a year.

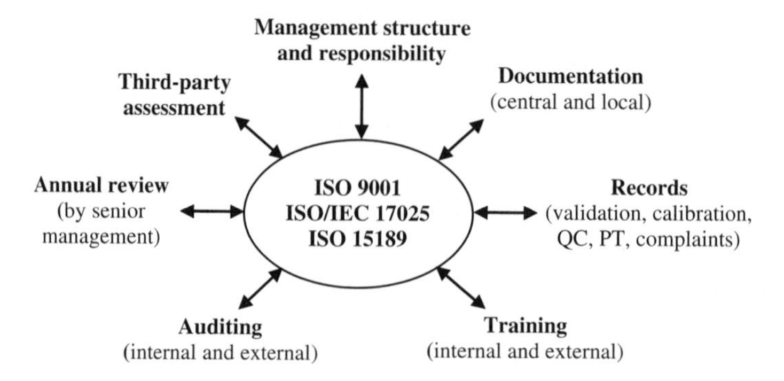

Figure 2.1 Main components of a quality management system.

2.3.2 Features of ISO 9001:2000

The quality management systems Standard, ISO 9001:2000, promotes the adoption of a process approach when developing, implementing and improving the effectiveness of a quality management system, to enhance customer satisfaction by meeting customer requirements. This is on the basis that any activity that uses resources to convert 'inputs' into 'outputs' can be considered to be a process. Figure 2.2 is based on Figure 1 from ISO 9001:2000, indicating how the Standard may be interpreted for a laboratory carrying out chemical analysis.

This Standard is a very general document and does not cover technical aspects. However, analytical laboratories increasingly have selected certification to ISO 9001 to cover the broader aspects of their activities. The earlier versions of this Standard had the term 'quality assurance' in the title but this has now been removed. The reason is that the current standard requires more than quality assurance of the service. Enhancement of customer satisfaction is also expected, along with evidence of continuous improvement. In Figure 2.2, this is covered in the box dealing with evaluating the results.

Changes take place in the laboratory system to take into account customer feedback and/or laboratory and office feedback. This could be after the completion of the work or as a result of the internal audit. Management review will evaluate the whole system and provide resource to enable changes to take place, e.g.

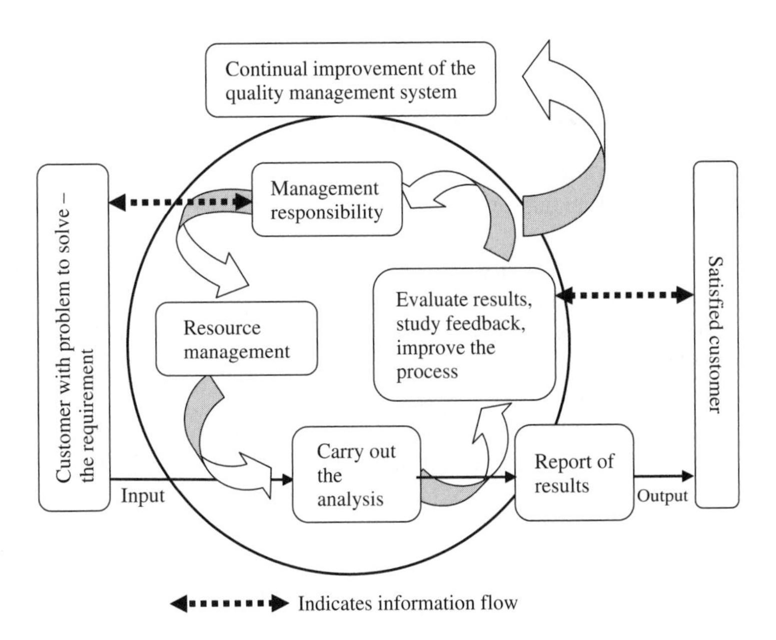

Figure 2.2 Model of an analytical problem in terms of a process-based management system.

extra/different equipment or staff. There is a great deal of emphasis on assessing the performance and having measures in place that demonstrate customer satisfaction.

2.3.3 Features of ISO/IEC 17025:2005

The ISO/IEC 17025 Standard, General Requirements for the Competence of Testing and Calibration Laboratories, is about competence and ways of demonstrating it. Whereas the ISO 9001 Standard is very general, ISO/IEC 17025 is very specific. Accreditation is given for specific tests in terms of the scope, i.e. the analyte, the matrix, method and concentration range. This Standard has two main sections – one dealing with the management requirements and the other with the technical requirements. To comply with the requirements of this Standard, the laboratory must operate a quality management system that meets the principles of ISO 9001. The technical section deals with the extra requirements of a laboratory wishing to demonstrate its competence in carrying out testing and/or calibration. Topics covered include method validation, measurement uncertainty and the use of reference materials and reference standards to achieve traceability of the reported results. Sampling that takes place in the laboratory is also covered along with the maintenance of sample integrity. As a means of assuring the quality of results, a laboratory is expected to participate in interlaboratory comparisons such as Proficiency Testing programmes (see Chapter 7).

2.3.4 Features of ISO 15189:2007

This Standard was prepared specifically to cover medical laboratories. These laboratories carry out testing or examination of materials derived from the human body. Such studies are carried out for the purpose of providing information for the diagnosis, prevention and treatment of disease or assessment of the health of an individual. ISO 15189 covers a whole range of tests relating to clinical measurements, e.g. chemical and microbiological. This Standard is based on ISO 9001 and ISO/IEC 17025 and contains a correlation between the clauses and those in ISO 9001 and ISO/IEC 17025.

The terminology used in this Standard is slightly different from the others in that it is appropriate for the particular discipline. For example, the term, 'referral laboratories' in paragraph 4.5 of ISO 15189 is used in a slightly different sense to the comparable clause in ISO/IEC 17025. Included in this section are consultants who may provide a second opinion. If the referral laboratory is an external laboratory, to which samples are submitted for a supplementary or confirmatory examination procedure and report, it is much the same as a 'contract laboratory' in ISO/IEC 17025. There is an extra Annex in this Standard which covers the ethics applicable to laboratory medicine.

2.3.5 Good Laboratory Practice (GLP)

The layout of the GLP principles is different from the Standards covered in the earlier sections but has much in common with the management sections within those Standards. It is narrow in its application but is a legal requirement for all laboratories carrying out regulatory work. Laboratories may wish to apply the principles if they are working in a different area but they are not eligible to register with the GLP Compliance Programme.

The aim of GLP is to encourage scientists to organize the performance of the study in such a way that ensures the production of reliable results. However, the GLP principles do not deal with the technical or scientific content of the research programme. They do not aim to evaluate the scientific value of the studies. However, it is expected that the methods used will be 'fit for purpose'. Use of standard methods of analysis is encouraged but the validation studies are not generally inspected during a quality assurance audit. It is important that the procedures used are fully documented, along with any changes made during the course of a study. Carrying out method validation under GLP compliance is not required by all monitoring authorities.

Emphasis is placed on the following organizational elements:

- resource, in terms of organization, personnel, facilities and equipment;
- written procedures and protocols;
- characterization of test items and test systems;
- documentation of raw data, final report and archiving;
- independent quality assurance unit;
- archiving.

It can be seen that the GLP principles do cover similar requirements to both ISO 9001 and ISO/IEC 17025 but data generated solely under these standards would not generally satisfy the principles of GLP. This is mainly in the area of documentation.

2.4 Best Practice

Implementing a quality management system for a laboratory is a formal way of implementing and demonstrating good laboratory practice or 'best practice' so as not to confuse it with GLP. It is possible for laboratories to achieve the implementation side, of the management standards already mentioned, by adopting some basic principles, without going for formal accreditation. In the UK, the Department of Trade and Industry (DTI) has funded research into achieving valid analytical measurements in the area of chemical measurement (and other areas of measurement) through the VAM programme [11]. Some years ago, an interdisciplinary group of measurement scientists suggested what are known as 'The six principles of valid analytical measurement'. These have subsequently been widely adopted and are given below. If you study the Principles and ISO/IEC 17025, you will see there is a great deal of common ground.

VAM Principles

1. Analytical measurements should be made to satisfy an agreed requirement.

   What does the customer want to know?

2. Analytical measurements should be made using methods and equipment which have been tested to ensure they are fit for purpose.

   Make sure everything works as it should and that the method used is capable of providing a useable result.

3. Staff making analytical measurements should be both qualified and competent to undertake the task.

   Use trained staff and do not use equipment unless you have been trained to use it competently.

4. There should be a regular and independent assessment of the technical performance of a laboratory.

   Take part in External Quality Assessment, formally through Proficiency Testing schemes and informally by taking part in intercomparison studies.

5. Analytical measurements made in one location should be consistent with those made elsewhere.

   Use Reference Materials so that your scale of measurement is acceptable at a national and international level.

6. Organizations making analytical measurements should have well defined Quality Control and Quality Assurance procedures.

   Implement a quality management system or at least the elements of such a system.

There are other ways of demonstrating that an organization implements best practice. One of these is obtained through the European Foundation for Quality Management (EFQM). The EFQM excellence model was introduced in 1992 as a framework for assessing applications for the European Quality Award which has since been replaced by the EFQM Excellence Award (EEA). The EFQM model provides a non-prescriptive framework currently based on eight fundamental concepts of excellence. These are as follows:

- **Results Orientation**

  Excellence is achieving results that delight all the organization's stakeholders.

- **Customer Focus**

  Excellence is creating sustainable customer value.

- **Leadership and Constancy of Purpose**

  Excellence is visionary and inspirational leadership, coupled with constancy of purpose.

- **Management by Processes and Facts**

  Excellence is managing the organization through a set of interdependent and interrelated systems, processes and facts.

- **People Development and Involvement**

  Excellence is maximizing the contribution of employees through their development and involvement.

- **Continuous Learning, Innovation and Improvement**

  Excellence is challenging the status quo and effecting change by utilizing learning to create innovation and improvement opportunities.

- **Partnership Development**

  Excellence is developing and maintaining value-adding partnerships.

- **Corporate Social Responsibility**

  Excellence is exceeding the minimum regulatory framework in which the organization operates and to strive to understand and respond to the expectations of their stakeholders in society.

More information can be obtained from the EFQM website [12].

Summary

This chapter outlines the means by which results which are fit for purpose are achieved. There are examples of how unreliable results can affect all of our lives. It explains some of the nomenclature encountered in quality management and why a quality management system is important. There is a brief description of the international standards that are applicable to a chemical analysis laboratory.

References

1. 'International Vocabulary of Metrology – Basic and General Concepts and Associated TErms (VIM)', ISO/IEC Guide 99:2007, International Organization for Standardization (ISO), Geneva, Switzerland, 2007.
2. 'Quality Management Systems – Fundamentals and Vocabulary', ISO 9000:2005, International Organization for Standardization (ISO), Geneva, Switzerland, 2005.
3. 'Quality Management Systems – Requirements', ISO 9001:2000, International Organization for Standardization (ISO), Geneva, Switzerland, 2000.
4. 'General Requirements for the Competence of Testing and Calibration Laboratories', ISO/IEC 17025:2005, International Organization for Standardization (ISO)/International Electrotechnical Commission (IEC), Geneva, Switzerland, 2005.
5. 'Conformity Assessment – Vocabulary and General Principles', ISO/IEC 17000:2004 International Organization for Standardization (ISO)/International Electrotechnical Commission (IEC), Geneva, Switzerland, 2004.
6. 'The Scope of Accreditation and Consideration of Methods and Criteria for the Assessment of the Scope in Testing', ILAC G18:2002, International Laboratory Accreditation Cooperation (ILAC), Silverwater, Australia, 2002.
7. 'UKAS Guidance on the Implementation and Management of Flexible Scopes of Accreditation within Laboratories', LAB39, United Kingdom Accreditation Service (UKAS), Feltham, UK, 2004.
8. 'Medical Laboratories – Particular Requirements for Quality and Competence', ISO 15189:2007, International Organization for Standardization (ISO), Geneva, Switzerland, 2007.
9. 'Good Laboratory Practice', Statutory Instrument 1999, No. 3106, Her Majesty's Stationery Office (HMSO), London, UK, 1999.
10. 'Good Laboratory Practice (Codification Amendments, etc.) Regulations 2004', Statutory Instrument 2004, No. 994, Her Majesty's Stationery Office (HMSO), London, UK, 2004.
11. Valid Analytical Measurement (VAM) Programme, LGC, Teddington, UK. [http://www.nmschembio.org.uk] (accessed 7 November, 2007).
12. European Foundation for Quality Management (EFQM), Brussels, Belgium. [http://www.efqm.org] (accessed 30 November, 2006).

Chapter 3
Sampling

Learning Objectives

- To understand the importance of sampling.
- To be able to identify different types of samples.
- To understand the importance of sampling plans and appreciate the legal and statutory requirements.
- To appreciate the importance of correct sample handling and storage.
- To be aware of the various problems associated with subsampling and the different techniques used.

This chapter gives an introduction to sampling. Devising a sampling plan or procedure may apply to your current position but it is more likely that the only sampling you are involved with is taking a test sample from the laboratory sample which has been submitted for analysis. However, you should be aware of the complete sampling process because this allows you to discuss sensibly the previous history of the material which comes into the laboratory. This will help to ensure that you measure the correct parameter.

The area of sampling is confused by the use of the same words in several different contexts. An IUPAC paper (1990) goes some way towards clarifying the definitions [1]. There is available also specific guidance on terminology in soil sampling [2]. In the foods sector, the Codex Alimentarius Commission has prepared draft general guidelines on sampling which were accepted by the Commission in July 2004 [3].

Quality Assurance in Analytical Chemistry E. Prichard and V. Barwick

3.1 Sampling Defined

Sampling is the process of selecting a portion of material, in some manner, to represent or provide information about a larger body of material. ISO/IEC 17025 defines sampling as follows:

> *A defined procedure whereby a part of a substance, material or product is taken to provide for testing or calibration a representative sample of the whole. Sampling may also be required by the appropriate specification for which the substance, material or product is to be tested or calibrated.*

The implications of the analysis have to be considered before taking the sample or devising a sampling scheme. It is the responsibility of the analytical chemist, through discussion with the customer, to establish the real nature of the problem. 'How much cadmium is there in this sample?' is not sufficiently specific. You must always ask why the information is required. The answer affects both the sampling plan and the choice of analytical method. These will depend on the acceptable level of uncertainty in the final result.

It is unfortunate that the sampling plan is often outside the control of the analyst. However, you should remember that while the analytical result *may* depend on the *method* used for the analysis, it will *always* depend on the type of *sampling plan* used. Knowledge of the potential uncertainty associated with sampling is important since if the sampling uncertainty is more than about two thirds of the total uncertainty, any attempt to reduce the analytical uncertainty is of little value. Sampling uncertainties cannot be evaluated or controlled using standards or reference materials. Evaluating the uncertainty associated with sampling can be complex; detailed guidance is generally 'sector-specific'. Appropriate guidance should be consulted if you need to estimate sampling uncertainties. However, if results for test samples are being reported on an 'as-received' basis, then the focus is only on the uncertainties associated with operations carried out within the laboratory (e.g. subsampling, sample pretreatment, measurement of the amount of analyte present). The topic of measurement uncertainty is discussed in detail in Chapter 6, Section 6.3.

DQ 3.1

What are the risks associated with poor sampling?

Answer

There are a number of risks associated with poor sampling. In general, poor sampling may result in the sample submitted to the laboratory for analysis being unrepresentative of the bulk material from which it was taken. This could result in substandard batches of material being accepted or perfectly good batches being rejected. In environmental analysis, poor sampling may result in invalid decisions being made about how to treat an area of contaminated land. This, in turn, could impact on public health. The exact nature of the risk depends on the reason for the sampling and subsequent analysis. You may have come up with slightly different examples but the ideas should be the same.

Figure 3.1 shows the relationship between the various operations in a sampling scheme and the analysis. This also helps identify some of the terms used. **Bulk materials** can take a number of forms. Examples include a single pile of a material, such as grain, soil present in an area of contaminated land or a shipload of coal. The key factor is that none of these examples are in the form of separate permanently identifiable units. In contrast, **packaged goods** are comprised of identifiable units, which may be assigned numbers. In some cases, bulk materials may be partially packaged into smaller units, such as bags or drums. These are called **segments**. A **consignment** (for both bulk materials and packaged goods) is defined as a quantity of material transferred on one occasion and covered by a single set of shipping documents. A **lot** is a quantity of material which is assumed to represent a single population for sampling purposes. A **batch** is a quantity of material which is known or assumed to have been produced under uniform conditions. A lot may consist of one or more batches and a consignment may, in turn, be made up of one or more lots. In the majority of cases, a lot or a batch is too large to allow a suitable laboratory sample to be obtained directly. There are, therefore, often a number of intermediate sampling stages required to obtain the sample which will be submitted to the laboratory. **Increments** are portions of the material obtained from the lot/batch by using a sampling device. Increments are often combined to produce a **primary** or **gross** sample. In some cases, the laboratory sample is obtained directly from the primary sample. In other situations, a number of primary samples are combined and mixed to produce a **composite** or **aggregate** sample. The laboratory sample is often obtained from the primary or composite sample by a series of division and reduction processes (e.g. coning and quartering, riffling). These processes are discussed briefly in Section 3.5.1. The **laboratory sample** is the portion of material delivered to the laboratory for analysis. The **test portion** is the quantity of material that is actually submitted for analysis. In the field of analytical chemistry, this is sometimes referred to as the **analytical portion**. If the laboratory sample is homogeneous, it may be possible to obtain the test portion directly, without further treatment of the sample. However, once the laboratory sample has been received, there are often a number of other operations that need to take place before a suitable test portion can be obtained. These are discussed in Section 3.5.1. The intermediate sample that the analyst obtains from the laboratory sample is called the **test** (or **analytical**) **sample**. The test portion is then taken from this test sample. The test portion itself often has to go through a number of treatment steps before the

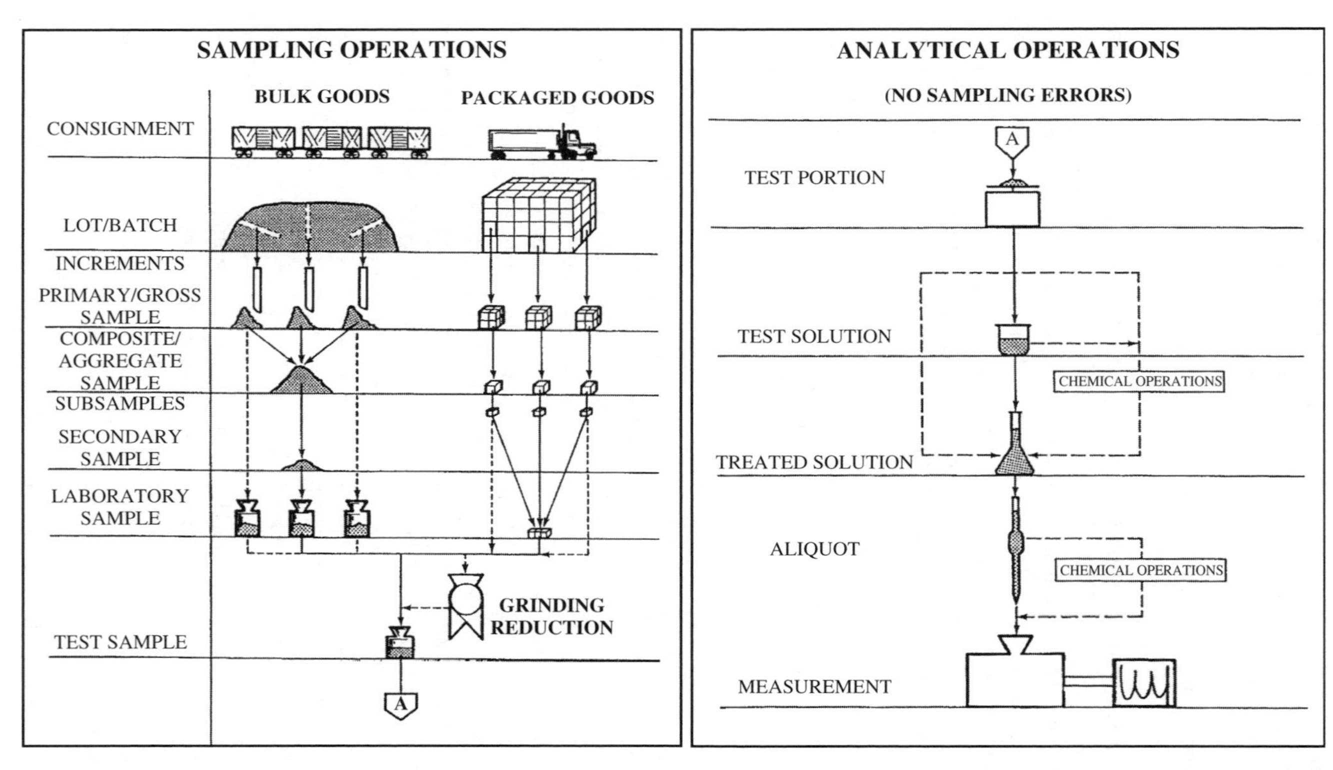

Figure 3.1 Schematic of sampling and analytical operations. Note: the lower "A" of the sampling operations continues with the upper "A" of the analytical operations [1]. Reproduced by permission of the International Union of Pure and Applied Chemistry, from Horwitz, W., *Pure Appl. Chem.*, **62**, 1193–1208 (1990).

final aliquot is obtained. The measurement of the property of interest is made on this aliquot. Some of the common sample-treatment procedures are discussed in Chapter 4, Section 4.5.4.

3.2 Types of Samples

The general definition of a sample is a 'portion of material selected from a larger quantity of material' [1]. However, there are several ways a sample can be described. Initially, a sample can be described in terms of its physical state (gas, liquid or solid). Where appropriate, these categories can be further subdivided into homogeneous or heterogeneous materials. A material may be described as heterogeneous because it can separate into more than one phase or, in the case of solid samples, because it contains a mixture of materials with varying particle sizes which may differ in composition. In the previous section, we saw that there are a number of different 'types' of sample (primary sample, laboratory sample, test sample, etc.) which will be encountered during the sampling process. Another way of describing samples is in terms of the sampling plan used to obtain the sample submitted to the laboratory. Using this descriptor, there are four types of samples: ***Representative***, ***Selective***, ***Random*** and ***Composite samples***.

3.2.1 Representative Sample

This is a sample that is typical of the parent material for the characteristic under inspection. You have to be careful in the way that you define the characteristic of interest. A sample may be adequate and representative if the concentration of the analyte is at a 5% mass/mass level (i.e. 5 parts per hundred) but it may not be acceptable if the analyte is present at the 5 mg kg^{-1} level (i.e. 5 parts per million). Knowledge of the method used for the analysis is also important. If the method produces results with an uncertainty of 30% (see Chapter 6, Section 6.3), the method of sampling need not be so finely controlled as in the case of a method which produces results with an uncertainty of only 5%.

To obtain an adequate representative sample we must take account of the state of the parent material we are to examine. There are four types, as follows:

(i) *Homogeneous* e.g. a vegetable oil at 40°C (at this temperature the oil is liquid); a filtered aqueous solution.

(ii) *Heterogeneous* e.g. palm oil at 15°C (this temperature is below the melting point of the oil); a sample of breakfast cereal, such as muesli.

(iii) *Static (contained) system* The composition of the parent material is permanent with respect to position in space and stable over the time of sampling and testing. There are many situations of this type, e.g. oil in a drum; tins of fruit in a warehouse.

(iv) *Dynamic system* The parent material is changing with respect to time. Removal of a portion at any instant represents only a 'snapshot' of that moment in time and in that particular location. The fact that it can never be reproduced presents difficulties in applying statistical control and consequently cannot be the subject of conventional statistical sampling plans, e.g. unsaturated and saturated oils being continuously blended; estuarine water, where the salinity is changing over time.

3.2.2 Selective Sample

This is a sample which is deliberately chosen by using a sampling plan that screens-out materials with certain characteristics and/or selects only material with other relevant characteristics. This may also be called *directed* or *focused* sampling.

DQ 3.2

Can you think of instances where this type of sample should be taken?

Answer

There are a number of situations where selective sampling would be appropriate. In food analysis, for example, it may be necessary to locate a specific adulterated portion of a lot, undiluted by perfectly good material. Other examples might be rodent contamination of flour by hair or urine, or toxic gases in a factory atmosphere where the total level may be acceptable but a localized sample may contain a harmful concentration.

3.2.3 Random Sample

A sample is selected by a random process to eliminate problems of bias in selection and/or to provide a basis for statistical interpretation of measurement data. There are three sampling processes which give rise to different types of random sample:

(i) *Simple random sampling.* Any sample has an equal chance of selection.

(ii) *Stratified random sampling.* The lot is subdivided/stratified and a simple random sample selected from each stratum.

(iii) *Systematic sampling.* The first sample is selected at random and then the subsequent samples are taken according to a previously arranged interval, e.g. every 5th, 10th or whatever is appropriate.

Each of the random samples obtained by the sampling processes described above has an equal chance of selection and so there is no bias. Note that a random sample may also be a representative sample, depending on the reason for the sampling and the nature of the material being sampled.

3.2.4 Composite Sample

Composite sampling is a way of reducing the cost of analysing large numbers of samples. A composite sample consists of two or more portions of material (collected at the same time) selected so as to represent the material being investigated. The ratio of components taken to make up the composite can be in terms of bulk, time or flow. The components of the composite sample are taken in proportion to the amount of the material that they represent. This type of sample may be appropriate when carrying out food surveys. The samples may, for example, be bulked in proportion to the amount normally consumed.

SAQ 3.1a

Sampling is not important because errors involved in sampling can be controlled by:

(i) use of standards true/false

(ii) use of reference materials true/false

SAQ 3.1b

Choose the most appropriate type of sample (representative, selective, random or composite) for the following parent materials.

(i) River water after a recent thaw. An estimate of the average concentration of compounds dissolved in the water is required.

(ii) Cans of baked beans in a warehouse.

(iii) Bars of chocolate suspected of being tampered with.

(iv) Sacks of flour stored near a hydrocarbon source in a ship's hold.

(v) Bags of flour in a storeroom, % moisture required.

3.3 The Sampling Plan

Sampling is always done for a specific purpose and this purpose will determine, to some extent, the sampling procedure used. Canned food is examined for leakage from the can, uniformity of contents and contamination. Crops need to be inspected during the growing season for levels of pesticides. Pharmaceutical products are examined for levels of active constituents and drug-release profiles. Regulatory samples of food are collected to determine if they conform to the label requirements and are safe to consume.

A **sampling plan** needs to be established which describes when, where and how samples are to be taken. IUPAC [1] define the sampling plan as follows:

> *A predetermined procedure for the selection, withdrawal, preservation, transportation and preparation of the portions to be removed from a population as samples.*

The definition shows that the sampling plan should include all aspects of the sampling process. It should include the number, location and size of the portions to be taken, and instructions for the extent of compositing and reduction of these portions to produce the laboratory sample. It should also address whether the process of sampling should be a 'one-off' or whether it should be repeated and if so, how often. When there is a regular requirement for analysis, the sampling plan is referred to as a **sampling scheme** or **sampling schedule**. The term **sampling programme** is often used to describe a combination of procedures where several related sampling schemes are combined.

The required sampling plan may be laid down in national or international standards or in a set of guidelines. Examples include the Codex Alimentarius Commission general guidelines on sampling [3], sampling procedures for monitoring water quality (e.g. ISO 5667-1) [4], sampling procedures for inspection by attributes (the ISO 2859 series of standards) [5–10] and sampling procedures for inspection by variables (ISO 3951) [11].

3.3.1 Legal and Statutory Requirements

There are regulations governing sampling schemes for a whole range of materials, e.g. for fertilizers and feeding stuffs. There are EC Directives which cover sampling, for example, the sampling of fruits and vegetables for examination for pesticide residues and for the determination of trace elements in fertilizers. At an international level, the Codex Alimentarius Commission has sampling schemes, e.g. for sampling foodstuffs for pesticide residues. You should familiarize yourself with any regulations dealing with your own area of work.

In general, the question you should always ask is, 'What will the results of the analysis be used for?'. If you are sampling for compliance with a contractual requirement, i.e. the sample must contain a minimum/maximum amount of the analyte, then it is important to know how this is interpreted. At the time of writing, the Codex Alimentarius Commission recommended the following limits for food-grade salt [12]:

- not more than 0.5 mg kg^{-1} arsenic;
- not more than 2 mg kg^{-1} copper;
- not more than 2 mg kg^{-1} lead.

One needs to know if this means, 'Individual items in a lot may not exceed ....' or 'The average of a number of items may not exceed ...'. You also need to know if the requirement is met if:

(a) a blended bulk sample could be formed from the sampled items

or

(b) each individual item is analysed and the average and distribution calculated.

Each of these interpretations requires a different approach. Note that blending material from a number of items prior to analysis (or averaging results obtained from a number of different items) may mean that 'hot spots' of the analyte in the sample are overlooked. This will be important if the purpose of the analysis is to study contamination which may affect only certain items (see Section 3.2.2).

There may also be cases where the maximum amount of analyte permitted is fixed by law, i.e. statutory limits. For these, there may be laid down standard procedures for sampling. The Codex maximum residue limit for the pesticide cypermethrin is 2 mg kg^{-1} (at the time of writing) in both citrus fruits and peaches [13]. When preparing the sample for analysis you need to know that in the case of the citrus fruit you take the whole fruit – skin, pith, pips, flesh and juice – whereas in the case of peaches it is the fruit after removal of stems and stones, but the residue is calculated and expressed in terms of the whole commodity (including the stone) without stems [14]. In some instances, it may be necessary to take a certain number of samples and that these must be taken in the presence of a witness.

3.3.2 Types of Sampling

3.3.2.1 Probability Sampling

Probability (or random) sampling allows a statistical evaluation to be applied to the data. It is used when a representative sample is required. There are three approaches which give rise to the three types of random sample described in Section 3.2.3.

Simple random sampling involves taking increments from the bulk material in such a way that any portion of the bulk has an equal probability of being sampled. This type of sampling is often used when little information is available about the material that is being sampled. It is also commonly used when sampling from a batch or consignment of manufactured product for quality control purposes.

Stratified sampling requires the consignment to be subdivided into groups (strata) according to predefined criteria. A simple random sampling plan is then applied to each stratum. The number of samples taken from each stratum is proportional to its size (e.g. weight or volume). The aim of stratified sampling is to produce a more representative sample than would be obtained by simple random sampling.

Systematic sampling is one of the most commonly used sampling techniques. This type of sampling involves taking increments from the bulk material at predetermined intervals, as defined by the sampling plan.

3.3.2.2 Non-Probability Sampling

This is used when a representative sample cannot be collected or is not appropriate. It is the correct sampling approach to use to produce a selective sample (see Section 3.2.2). There are three main non-probability sampling strategies.

Judgement sampling involves using knowledge about the material to be sampled, and the reason for sampling, to select specific samples.

Quota sampling requires the consignment to be subdivided into groups (as for stratified sampling described previously). Once the material has been grouped, judgement sampling is used to select samples from each group.

Convenience sampling involves selecting samples on the basis of availability and/or accessibility.

3.3.2.3 Bulk Sampling

This type of sampling involves the taking of a sample from material which does not consist of discrete, identifiable or constant units. The bulk material may be gaseous, liquid or solid.

3.3.2.4 Acceptance Sampling

Acceptance sampling involves the application of a predetermined sampling plan to decide whether a batch of goods meets the defined criteria for acceptance. The main aim of any acceptance sampling must be to see that the customer gets the quality required, while remembering that financial resources are limited and that the cost of the article must reflect the cost of inspection, as well as the cost of production.

Acceptance sampling can be either **by attributes** or **by variables**. In sampling by attributes, the item in the batch of product either conforms or not. The number of nonconformities in the batch is counted and if this reaches a predetermined figure, the batch is rejected. In sampling by variables, the characteristic of interest is measured on a continuous scale. If the average meets a predetermined value, and the variation in the characteristic being measured is within an acceptable standard deviation, the batch is accepted.

To illustrate the difference between these two types of sampling plan, let us look at an example. Cornflakes are sold in packets of 500 g. In attributes sampling, each packet that weighs 500 g or more is accepted, and each packet that weighs less than 500 g is rejected. If the number of rejects is less than the predetermined number, the batch is accepted. If you have to sample by variables, the packets are weighed and the actual weights are averaged and the standard deviation of the weights calculated. If the mean weight meets or exceeds the declared average and the magnitude of the standard deviation does not indicate any unreasonable shortages, the batch is accepted.

There are international standards detailing the sampling procedures for both of these approaches [5–11].

3.4 Sample Numbers and Sample Size

The sampling plan should specify the number and size of primary samples which need to be obtained from the lot/batch. It should also describe how the laboratory sample is to be obtained. These issues may well be outside of the analyst's control, but it is important to consider how the validity of any analysis will be affected.

Most chemical tests are destructive and so all the material cannot be tested. In any case, this would not be very cost-effective. There may be a problem in taking a representative sample from bulk material known to be heterogeneous. The sampling plan must be such that the degree of homogeneity can be tested.

If the validated test method requires 1 g of material but only 100 mg is available, you must find out if the method is sufficiently robust to stand this amount of scaling down. This has to be checked **before** the analysis starts, i.e. the method must be validated for analysis of 100 mg of material. Even if the method of analysis is found to be robust, scaling down is only a viable option if the smaller test portion size remains representative, within acceptable limits. This will depend on the homogeneity of the material.

In addition to the sampling that goes on external to the laboratory, the analyst must also address issues such as the size and number of test portions which will be taken from the laboratory sample for analysis.

We have to be careful when talking about 'sample size' and be sure that we know the context in which it is being used. 'Sample size' is sometimes used to describe the **number** of samples taken from a larger amount of material, such as a lot or a batch. This is a statistical term used to describe the number of 'items' selected from a larger population. This will be specified in the sampling plan. However, in the laboratory, an analyst may use 'sample size' to describe the amount of material that forms the test sample or test portion. Both meanings are important and need to be addressed when considering sampling, although the former is often outside the control of the analyst. You can avoid this confusion by using the correct terminology, e.g. size of test sample, size of test portion, number of primary samples, etc.

3.4.1 Sampling Uncertainty

In order to determine how many samples we require, it is necessary to consider the sources of *uncertainty* in the final result. Uncertainty is dealt with in more detail in Chapter 6. In this section, we are mainly concerned with the uncertainty arising from sampling. It is necessary to use a few statistical terms namely, sample standard deviation (s) and variance (s^2). These terms are defined in Chapter 6, Section 6.1.3.

The total variance in the final result (s^2_{total}) is made up of two contributions. One is from variation in the composition of the laboratory samples due to the nature of the bulk material and the sampling procedures used (s^2_{sample}). The other (s^2_{analysis}) is from the analysis of the sample carried out in the laboratory:

$$s^2_{\text{total}} = s^2_{\text{sample}} + s^2_{\text{analysis}} \tag{3.1}$$

The analytical variance can be determined by carrying out replicate analysis of samples that are known to be homogeneous. You can then determine the total variance. To do this, take a minimum of seven laboratory samples and analyse each of them (note that s^2_{sample} characterizes the uncertainty associated with producing the laboratory sample, whereas s^2_{analysis} will take into account any sample treatment required in the laboratory to obtain the test sample). Calculate the variance of the results obtained. This represents s^2_{total} as it includes the variation in results due to the analytical process, plus any additional variation due to the sampling procedures used to produce the laboratory samples and the distribution of the analyte in the bulk material.

The sample variance is then given by:

$$s^2_{\text{sample}} = s^2_{\text{total}} - s^2_{\text{analysis}} \tag{3.2}$$

The variance of the sample (s^2_{sample}) is also made up of two components, i.e. that due to the population, s^2_{pop} (i.e. the variation of the distribution of the analyte throughout the material) and that due to the sampling process (s^2_{slg}). You should always try to make sure the variance due to sampling is negligible. The variance due to the population is the one that is of most concern to the analyst:

$$s^2_{\text{sample}} = s^2_{\text{pop}} + s^2_{\text{slg}} \tag{3.3}$$

The magnitude of each of these components will influence the number of samples you need to take so as to achieve a given overall uncertainty.

3.4.2 Number of Primary Samples

Each sector has specific requirements in terms of the number of primary samples that should be taken. Simple empirical rules which have been used in the past to determine the number of samples to be taken from a lot include $n = \sqrt{N}$ and $n = 3 \times \sqrt[3]{N}$ (in each case, N is the total number of items in the lot). In both cases, n is rounded to the nearest integer.

In addition, published sampling plans for different sectors indicate the number of samples to be taken from a lot. For example, when carrying out acceptance sampling by attributes (see Section 3.3.2), as described in ISO 2859-1 [6], a number of factors have to be taken into account:

- *Lot/batch size*

This needs to be known as it will influence the number of samples that need to be taken from the lot.

- *Inspection level*

This relates the number of samples to the size of the lot and therefore determines the ability to discriminate between good and poor quality lots. The inspection level to be used for a particular requirement will be prescribed by the relevant responsible authority. Generally, there are three possible inspection levels, i.e. Levels I, II and III. Typically, Level II inspection is used unless another inspection level is specified. Level I is used when less discrimination is required, whereas Level III is used when greater discrimination is required. ISO 2859-1 contains a table which allows the user to identify the appropriate sample size, depending on the size of the lot and level of inspection required. This is reproduced in Table 3.1. Each letter refers to a sampling plan in the standard (see Table 3.2 for an example of a sampling plan). You will see that in addition to the general inspection levels I, II and III, there are four special inspection levels (S1–S4). These are used when the sample size must be kept small and larger sampling risks can be tolerated.

Once the lot size and inspection level are known, Table 3.2 can be used to identify the appropriate sampling plan. For example, if the lot size is 4000 and inspection level II is required, then L is the appropriate code letter. The code letter relates to the number of items (samples) from the lot that needs to be examined, as shown in Table 3.2.

- *Type of sampling*

Single, double or multiple sampling may be used. The nature of the analysis required will determine which one is used. Single sampling means that an appropriate number of items are taken from a batch and they are investigated. Double sampling is a system in which a first group of items is taken that is smaller (i.e. contains fewer items) than would be taken for single sampling. If the quality is

Table 3.1 Sample size code letters (ISO 2859-1). When linked with a particular sampling plan, the code letter indicates the required sample size [6][†]

Lot or batch size	Special inspection levels				General inspection levels		
	S1	S2	S3	S4	I	II	III
2 to 8	A	A	A	A	A	A	B
9 to 15	A	A	A	A	A	B	C
16 to 25	A	A	B	B	B	C	D
26 to 50	A	B	B	C	C	D	E
51 to 90	B	B	C	C	C	E	F
91 to 150	B	B	C	D	D	F	G
151 to 280	B	C	D	E	E	G	H
281 to 500	B	C	D	E	F	H	J
501 to 1200	C	C	E	F	G	J	K
1201 to 3200	C	D	E	G	H	K	L
3201 to 10 000	C	D	F	G	J	L	M
10 001 to 35 000	C	D	F	H	K	M	N
35 001 to 150 000	D	E	G	J	L	N	P
150 001 to 5 000 000	D	E	G	J	M	P	Q
5 000 001 and over	D	E	H	K	N	Q	R

[†]The terms and definitions taken from ISO 2859-1:1999 Sampling procedures for inspection by attributes Part 1: Sampling schemes indexed by acceptance quality limit (AQL) for lot-by-lot inspection, Table 1, a portion of Table 2-A, and a portion of Table 10-L-1, are reproduced with permission of the International Organization for Standardization, ISO. This standard can be obtained from any ISO member and from the Web site of ISO Central Secretariat at the following address: www.iso.org. Copyright remains with ISO.

found to be sufficiently good, the batch may be accepted. If the quality is sufficiently bad, the batch may be rejected. If the first group of items is marginal in quality, then the second group of items is taken and examined before a decision is made (normally the first and second groups contain the same number of items). Multiple sampling is, in principle, the same as double sampling but more than two groups of items are taken. The operating characteristics curves (see later) for all three types are almost identical and therefore the proportion of batches accepted would be almost the same whichever is used. Therefore, we shall only deal with single sampling.

- *Inspection type*

Normal, tightened or reduced inspection may be used. Tightened inspection is introduced when two out of five consecutive lots have to be rejected. This is normally kept in force until the inspector is satisfied that quality has been restored. If the quality is consistently better than the Acceptance Quality Limit (AQL), then reduced inspection can be introduced which requires two fifths of the number of samples taken for normal inspection.

Table 3.2 Single sampling plans for normal inspection (ISO 2859-1) [6][†]

Sample size code letter[a]	Sample size[b]	Acceptance Quality Limit in percent nonconforming items and nonconformities per 100 items (normal inspection)[c]																					
		0.1		0.15		0.25		0.4		0.65		1.0		1.5		2.5		4.0		6.5		10	
		Ac	Re	Ac	Re	Ac	Re	Ac	Re	Ac	Re	Ac	Re	Ac	Re	Ac	Re	Ac	Re	Ac	Re	Ac	Re
F	20	⇓		⇓		⇓		⇓		0	1	⇑		⇓		1	2	2	3	3	4	5	6
G	32	⇓		⇓		⇓		0	1	⇑		⇓		1	2	2	3	3	4	5	6	7	8
H	50	⇓		⇓		0	1	⇑		⇓		1	2	2	3	3	4	5	6	7	8	10	11
J	80	⇓		0	1	⇑		⇓		1	2	2	3	3	4	5	6	7	8	10	11	14	15
K	125	0	1	⇑		⇓		1	2	2	3	3	4	5	6	7	8	10	11	14	15	21	22
L	200	⇑		⇓		1	2	2	3	3	4	5	6	7	8	10	11	14	15	21	22	⇑	
M	315	⇓		1	2	2	3	3	4	5	6	7	8	10	11	14	15	21	22	⇑		⇑	
N	500	1	2	2	3	3	4	5	6	7	8	10	11	14	15	21	22	⇑		⇑		⇑	
P	800	2	3	3	4	5	6	7	8	10	11	14	15	21	22	⇑		⇑		⇑		⇑	
Q	1250	3	4	5	6	7	8	10	11	14	15	21	22	⇑		⇑		⇑		⇑		⇑	
R	2000	5	6	7	8	10	11	14	15	21	22	⇑		⇑		⇑		⇑		⇑		⇑	

[a]The sample size code letter is obtained from Table 3.1.
[b]The sample size indicates the number of items that need to be selected from the lot for examination.
[c]⇓, use first sampling plan below arrow; ⇑, use first sampling plan above arrow; Ac, acceptance number; Re, rejection number.
[†]The terms and definitions taken from ISO 2859-1:1999 Sampling procedures for inspection by attributes Part 1: Sampling schemes indexed by acceptance quality limit (AQL) for lot-by-lot inspection, Table 1, a portion of Table 2-A, and a portion of Table 10-L-1, are reproduced with permission of the International Organization for Standardization, ISO. This standard can be obtained from any ISO member and from the Web site of ISO Central Secretariat at the following address: www.iso.org. Copyright remains with ISO.

- *Percent nonconforming items*

This is the ratio of the number of nonconforming items (samples) to the number of items examined, multiplied by 100. Note that if each item is being tested for more than one analyte, then it is appropriate to use 'nonconformities per 100 items' rather than 'percent nonconforming items'. For example, if a consignment of baked beans is being analysed for protein, carbohydrate and fat content, it is possible that an item could fail on more than one of the analytes. If 200 cans are examined and it is found that three are low in protein, two are low in carbohydrate and protein, and one is low in fat, protein and carbohydrate, there are six nonconforming items but a total of ten nonconformities. The percent nonconforming items is given by:

$$\frac{6}{200} \times 100 = 3$$

while the number of nonconformities per 100 items is given by:

$$\frac{10}{200} \times 100 = 5$$

- *Acceptance quality limit* (AQL)

The AQL is related to the quality required in the product. It is defined in ISO 2859-1 as 'The quality level that is the worst tolerable process average when a continuing series of lots is submitted for acceptance sampling'. It is, therefore, the maximum percent nonconforming items that, for the purpose of the sampling inspection, can be considered acceptable as a process average. However, the AQL should not be taken as a 'target' level for nonconforming items. In fact, ISO 2859 is designed to encourage manufacturers to have process averages that are consistently better than the AQL – otherwise there is a risk of switching to tighter inspection (see above).

- *Acceptance number* (Ac)

For a given sampling plan, the acceptance number is the maximum number of nonconforming items allowed in the group of items selected for inspection, if the lot is to be accepted.

Table 3.2 shows a summary of single sampling plans for normal inspection (note: not all sample size code letters or AQLs are shown – please refer to ISO 2859-1 for a complete set of sampling plans). The sample code tells you the

Table 3.3 Tabulated operating characteristics, single sampling plans, normal inspection level with $n = 200$ [6][†]

	Acceptance Quality Limit (normal inspection)		
	0.65	2.5	6.5
P_a [a]		p [b]	
99.0	0.414	2.42	6.43
95.0	0.686	3.11	7.57
90.0	0.875	3.54	8.22
75.0	1.27	4.33	9.40
50.0	1.83	5.33	10.8
25.0	2.54	6.46	12.4
10.0	3.31	7.60	13.8
5.0	3.83	8.33	14.8
1.0	4.93	9.82	16.6

[a] Percentage of lots expected to be accepted.
[b] Percent nonconforming items.
[†] The terms and definitions taken from ISO 2859-1:1999 Sampling procedures for inspection by attributes Part 1: Sampling schemes indexed by acceptance quality limit (AQL) for lot-by-lot inspection, Table 1, a portion of Table 2-A, and a portion of Table 10-L-1, are reproduced with permission of the International Organization for Standardization, ISO. This standard can be obtained from any ISO member and from the Web site of ISO Central Secretariat at the following address: www.iso.org. Copyright remains with ISO.

number of items (i.e. size of sample) that need to be taken from the lot for examination. For example, code L requires a sample containing 200 items. You also need to know the AQL. Once you know this, the table can be used to determine the maximum number of nonconforming items (or number of nonconformities) permitted in the group of items examined. If the AQL is 2.5%, then if there are 10 or fewer nonconforming items the batch is accepted. This is the acceptance number (Ac). If there are 11 or more nonconforming items then the batch may be rejected. This is the rejection number (Re).

Each sampling plan should be accompanied by an **operating characteristics curve** (or OC curve). The curve describes the probability of acceptance of a lot as a function of its actual quality. It shows what any particular sampling plan can be expected to achieve in terms of accepting or rejecting lots.

Table 3.3 shows the operating characteristics for a normal inspection level sampling plan with a sample size (i.e. number of items examined) of $n = 200$ at three different AQLs. This table shows that if a product has 3% nonconforming items (p) then, using an AQL of 2.5%, approximately 95% of lots would be expected to be accepted. Operating characteristics curves are given in ISO 2859-1.

SAQ 3.2

A sampling plan is required for the purpose of inspecting the quality of bags of frozen peas. A bag of peas is considered unsatisfactory (i.e. to be a nonconforming item) if it contains more than 10 wt% defective peas (e.g. blemished peas, blond peas, etc). The bags of peas are produced in lots consisting of 3000 bags of peas. Level II inspection is required, the inspection type is normal and the AQL has been set at 6.5%. Use the information given in Table 3.1 and Table 3.2 to determine:

- the number of samples required (i.e. the number of bags of peas that must be selected from the lot at random for testing) assuming a single sampling plan;
- the maximum number of the sampled bags that can be nonconforming (i.e. contain more than 10 wt% defective peas) if the lot is to be accepted.

3.5 Subsampling

A subsample is a portion of a sample, prepared in such a way that there is some confidence that it has the same concentration of analyte as that in the original sample. The laboratory sample may be a subsample of a bulk sample and a test sample may be a subsample of the laboratory sample. Because of inhomogeneity, differences may occur between samples but there should not be any significant inhomogeneity between subsamples.

Although the error associated with subsampling carried out in the laboratory to obtain the test portion is sometimes insignificant, it can be much greater than intuition would predict. It becomes more important as the concentration of the analyte of interest diminishes. In analyses for trace elements, it probably constitutes one of the largest single sources of experimental error. The size of a test portion is often dictated by the method used and can range from a gram to micrograms.

An indication of the minimum size of a subsample can be obtained by using the concept of a sampling constant. For example, in the laboratory, the sampling constant can be used to estimate the minimum size of the test portion. However, the suitability of the chosen test portion size must be confirmed as part of method validation. The sampling constant K_s has units of mass. This is the mass of the test portion necessary to ensure a relative subsampling error of 1% (at the 68% confidence level) in a single determination. The value of $\sqrt{K_s}$ is numerically equal to the coefficient of variation, CV (see Chapter 6, Section 6.1.3) for results obtained on 1 g subsamples in a procedure with insignificant analytical error.

If the laboratory sample has been prepared in a particular way to pass a specific mesh size, the coefficient of variation of the result for one component varies inversely with $\sqrt{m}$, where m is the mass of the test portion. A sampling constant (K_s) can be defined by the following:

$$CV = \sqrt{\frac{K_s}{m}} \text{ or } K_s = (CV)^2 m \tag{3.4}$$

This relationship presumes that the test portion corresponds to at least a certain minimum number of particles and that the sample is 'well-mixed'. The combined results from two test portions, each of mass m, has the same subsampling variability as a single test portion of mass $2m$.

To determine the value of CV, you need to analyse a number of equally sized test portions of a well-mixed material, each of mass m. Calculate the CV and hence determine K_s.

The estimate of K_s can be used as follows:

(i) If the target CV for the method is known, the test portion size (m_i) required to achieve that CV can be calculated by using equation (3.5):

$$m_i = \frac{K_s}{(CV)^2} \tag{3.5}$$

(ii) Once K_s has been evaluated for an analyte in a particular sample type, the coefficient of variation (CV_f) for the same analyte in a future test portion of mass m_f is then estimated by:

$$(CV)_f = \sqrt{\frac{K_s}{m_f}} \tag{3.6}$$

3.5.1 Subsampling Procedures

As mentioned previously, subsampling can occur both outside of the laboratory (to obtain a laboratory sample from the lot/batch) and within the laboratory (to obtain a test portion from the laboratory sample). In both cases, it is important that suitable subsampling procedures are used to ensure that the subsample is representative of the larger amount of material. There are many different operations that can be carried out to obtain suitable samples for analysis. Some examples for different types of material are given below. There may be legislation governing the subsampling protocol that you should use for your analysis. You must consult the appropriate documentation for this.

3.5.1.1 Solid Material

Sampling of solid material often produces much more material than can be submitted to the laboratory for analysis. Reducing a large composite sample to a suitably sized laboratory sample often involves the following three steps:

- milling/grinding by mechanical means to produce a mixture containing particles of the appropriate size;
- mixing/homogenization by using a ball mill;
- subdivision of the ground and mixed sample using coning and quartering or riffling techniques.

'Coning and quartering' is used to reduce the size of granular and powdered samples. The sample is placed on a flat surface in the form of a conical heap. The heap is then spread out and flattened into a circular cake, which is then divided into approximately equal quarters. One pair of opposite quarters is removed, combined and formed into a new cone for the process to be repeated (the other two quarters are discarded). The process is repeated as many times as is necessary to obtain a sample of the required size.

A 'riffler' is a mechanical device, consisting of a metal box containing a number of equally spaced slots, which is used for dividing a sample into two approximately equal portions. The material to be subdivided is poured into the top of the box and emerges through the slots on opposite sides in two approximately equal portions. As with coning and quartering, the procedure is repeated until the desired sample size is obtained.

The analyst in the laboratory may also have to carry out one of the above procedures to obtain a suitable test portion from the laboratory sample. Prior to taking the test portion for analysis, the test sample may require additional treatment.

'Defatting' is an example of a commonly used sample pretreatment process where the analyte is insoluble in non-polar solvents. Lipids interfere in many analytical processes and so are removed at the start of the analysis, if possible, by washing with a non-polar solvent, such as hexane. Removal of fat may also assist in subsequent solvent penetration for extraction of the analyte.

Extraction of an analyte from a complex matrix, such as foods, is often dependent on the moisture and lipid content of the matrix. Hence, sample pretreatment may involve drying (to remove excess moisture) or rehydration under controlled conditions of relative humidity to improve solvent penetration.

The addition of water or the use of an aqueous solvent mixture is important for the extraction of other organic analytes from dry foodstuffs or dehydrated foods. It is particularly necessary in aiding the permeation of solvent through freeze-dried samples.

Soils have to be dried under controlled conditions so as to avoid changing their chemical composition. You should remember that microbial activity can affect the levels of some analytes, such as phosphorus and potassium.

3.5.1.2 Liquid Material

Subsampling of liquids may appear to present much less of a problem than solid samples. However, this is only the case when the volume of liquid to be sampled is small enough that it can be homogenized by shaking, and the liquid consists of only one phase.

Liquids often contain sediment or other solid matter in suspension. The presence of suspended material can affect the determination of the concentration of the analyte. The suspended material may adsorb the analyte and so it is important to check whether filtration, if used, has a significant effect on the analytical result.

In some cases, the analyte may be in suspension rather than in solution in the test sample, e.g. metals in engine oil. In such cases, it is important that the bulk sample is adequately mixed (homogenized) before a subsample is taken, as sedimentation may have occurred. If the sample is not adequately mixed, then the result obtained for a sample taken from the bulk will give a biased estimate of the true value of the analyte. Liquids may also settle in layers on standing. It is important that there is sufficient 'headspace' in the container for adequate shaking. When material is prone to rapid sedimentation, the samples need to be taken *during* the mixing process as the material will immediately start to separate once the mixing is stopped.

SAQ 3.3

Would you homogenize the contents of the following cans before analysing for trace elements?

(i) Canned tuna in brine.

(ii) Canned peaches in syrup.

(iii) Canned grapefruit in natural juice.

(iv) Canned fish in tomato sauce.

3.6 Sample Handling and Storage

When a sample is received it should have a unique identification, i.e. a number or code. All details about the sample should be recorded. This will include storage conditions and, if it is necessary to transfer the sample from person to person, this should be fully documented. Details of the container and closures should also be recorded. These may have been inappropriate and influence the analytical result. The appearance of the sample on receipt should also be documented.

Ideally, you should examine the sample as soon as possible after receipt, provided that the scope of the analysis and the methods to be used are clear and have been agreed. Storage conditions and the length of storage should be recorded.

Properties of the analyte, such as volatility, sensitivity to light, thermal stability and chemical reactivity, all have to be considered when designing a sampling strategy. These factors need to be taken into account to ensure the quality of the sample does not degrade before the measurements are made.

The samples should be stored so that there is no hazard to laboratory staff. The integrity of the sample must also be preserved, i.e. the sample should be the same when it is analysed as when it was collected. There must be no risk of contamination or 'cross-contamination', i.e. no material should enter or leave the sample container. In addition, extremes of environmental conditions should be avoided.

For trace level analysis, you must store samples in a separate area away from analytical calibrants or any other material which may contain a high concentration of the analyte. Storage is usually in a completely segregated and dedicated room. It may be necessary to take precautions to avoid cross-contamination between sample storage areas and other laboratory areas. This can include changing laboratory coats when entering and leaving the storage area and the use of disposable adhesive mats to prevent any material being 'walked' between rooms.

Conditions of storage prior to analysis should be agreed in advance with the customer. This may involve storage in a cupboard, storeroom, refrigerator, freezer or cold room as appropriate. The choice will depend on the properties of the sample and the need to protect the sample from light, elevated temperature or humidity (see Table 3.4). You may need to use a maximum/minimum thermometer to check for temperature fluctuations during storage. The samples must also be stored under appropriate conditions during the time interval between sampling and arrival at the laboratory for analysis.

Most analytes and sample matrices are more stable at low temperatures and so freezing the sample is the usual 'first-choice' method of storage. At deep-freeze temperatures (-18°C), most enzymatic and oxidative reactions are reduced to a minimum. However, changes in biological samples can occur during freezing and thawing as these processes disrupt the cellular structure. Some fruits and vegetables soften and blacken upon thawing. Fluid may be released from ruptured cells which, if ignored, can reduce sample homogeneity. In addition, samples which are emulsions should not be frozen. If a number of test portions are to

Table 3.4 Storage conditions for laboratory samples

Storage condition	Appropriate sample types	Inappropriate sample types
Deep freeze (−18°C)	Samples with high enzymatic activity Perishable goods/products Less stable analytes	Samples which liquefy on thawing Aqueous samples
Refrigerator (4°C)	Soils Fresh fruit and vegetables Aqueous samples	Samples with possible enzymatic activity
Room temperature (in the dark)	Dry powders and granules Minerals Stable analytes	Fresh foods
Desiccator	Hygroscopic samples	Samples which are more hygroscopic than the desiccant

be analysed, it is often convenient to homogenize and subdivide the laboratory sample prior to freezing. The test portions can then be kept in the freezer until they are required. This approach is particularly useful if a number of tests need to be carried out over a period of time – only one portion of sample has to be removed from the freezer at any one time for a particular test, so that the laboratory sample is not subjected to repeated thawing and freezing cycles which could cause deterioration of the sample.

Samples which cannot be frozen or which do not need to be frozen, such as soils intended for elemental analysis, plastics, paints or any other samples where both the matrix and the analyte are non-volatile and stable at ambient temperatures, are usually stored at 0 to 5°C. It may not even be necessary to refrigerate the sample, as many are stable at room temperature.

There are various physical and chemical methods of arresting or slowing down sample degradation (see Table 3.5). It is vital to verify that the integrity of the analyte is not affected by the method used to prevent degradation. This should be checked during method validation.

Freeze-drying can be a useful method of preserving friable samples with a moderate moisture content (such as breadcrumbs) and is also an effective way of preconcentrating aqueous samples. It is not appropriate for volatile analytes, e.g. mercury, as there are likely to be significant losses during the freeze-drying process. Irradiation of samples for long-term storage is used particularly where it is desirable to minimize bacteriological activity in a sample, for example, to inhibit the growth of moulds in water samples. Antioxidants can be added to liquids and

Table 3.5 Examples of physical and chemical methods of preserving samples

Method	Examples of applications
Freeze-drying[a]	Breads, biscuits, etc., aqueous samples
Irradiation[b]	Aqueous samples, biological samples
Adding antioxidants[b,c]	Liquids and solutions
Adding anticoagulants[c]	Blood and clinical samples
Autoclaving[b]	Sterilizing body fluids

[a]Unsuitable for volatile analytes.
[b]Stability of analyte must be established.
[c]Check specific interference effects.

solutions to prolong the stability of unstable analytes, such as vitamins or unstable matrices such as vegetable oils. Anticoagulants, such as heparin and ethylenediamine tetracetic acid (EDTA), are frequently added to blood samples to prevent clotting and thus preserve the sample in a form suitable for analysis. The type and amount of anticoagulant used will depend on the type of analysis required – care must be taken to select an anticoagulant which will not interfere with the analysis.

Precautions may also have to be taken to prevent loss or gain of moisture, and to prevent photochemical degradation. Light-sensitive samples should be stored in the dark, in amber glass containers or in glass containers protected by aluminium foil. Samples containing volatile constituents should be kept in well-sealed containers and preferably stored in the cold to reduce the vapour pressure of such compounds.

All samples should normally be allowed to reach ambient temperature before analysis. Care should be taken to avoid hygroscopic samples taking up water, both when stored in a deep freeze or refrigerator and when warming back to room temperature.

Thought has to be given to the appropriate type of container, closure and label before setting out to collect the sample. Glass may be thought of as an inert material but it is not suitable for some samples.

DQ 3.3

Can you think of examples where glass would not be a suitable container for samples?

Answer

Glass containers may adsorb or desorb elements. Sodium can desorb from soft glass and borosilicate glass but soft glass is a more serious problem when analysing trace levels of inorganic materials. Glass containers are often cleaned by using phosphate detergents and even after washing with acid and several rinses with water, high phosphorus levels are recorded. So, for many trace analyses, glass may not be suitable.

> Storing aqueous samples in glass containers prior to being examined for polynuclear aromatic hydrocarbons at the ng l^{-1} level can also be a problem. There is evidence of adsorption of the hydrocarbons onto the glass surface. This problem is considerably reduced by putting the extracting solvent in the bottle before introducing the water sample.

Polyethylene is another common container material. Polyethylene bottles are suitable for most solids and aqueous samples. When used for aqueous samples, unlike glass there will be no leaching of elements such as Na, K, B and Si. For best results when determining metal ions in aqueous samples, the sample should be acidified to avoid precipitation of the ions.

Some samples may change on standing. For example, the cream separates out from milk samples and the buttery lumps have to be broken up before the analysis. The composition of other samples may change due to, for example, fermentation.

Ideally there should be sufficient sample for visual examination. The observation should be recorded and any change that takes place should also be noted.

3.6.1 Holding Time

It is essential that the sample on which the final measurement is made has the *same* composition as the material at the time of sampling. Holding time is defined as the maximum period of time that can pass from sampling to measurement before the sample has changed significantly. Holding time is important when considering storage. When degradation is possible, samples should be measured before any significant change has occurred. To calculate storage time, a large sample is taken and this is stored under normal conditions. Test portions are withdrawn at regular intervals and measured in duplicate. This allows an estimate of the sample standard deviation (s) to be obtained as follows:

$$s = \sqrt{\frac{\sum_{i=1}^{n} d_i^2}{2n}} \tag{3.7}$$

where d_i is the difference between pairs of duplicates and n is the number of pairs of duplicates.

The mean values of the duplicate measurements are plotted with respect to time and the best-fit straight line drawn through the points. The point at which the line reaches a value of $3s$ less than the initial value gives the maximum holding time. Figure 3.2 shows a graph of this type.

If the holding time is inconveniently short, then the storage conditions have to be changed or the sample may have to be stabilized in some way.

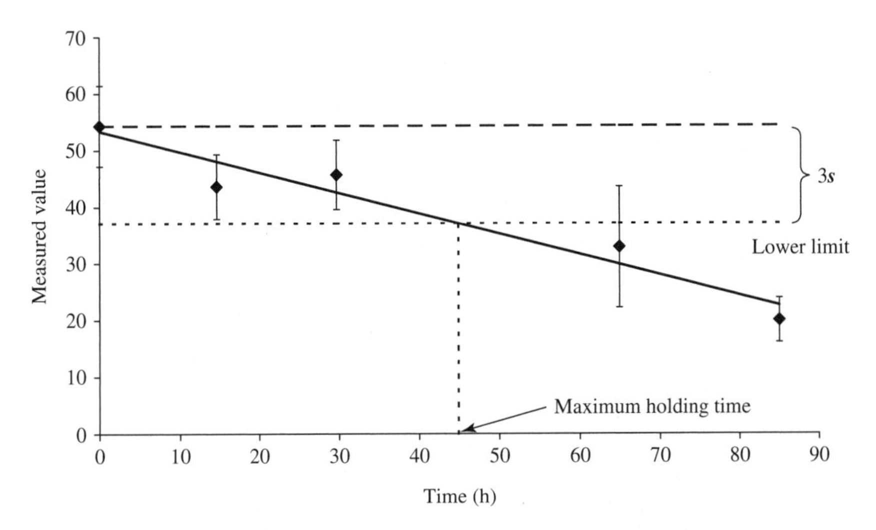

Figure 3.2 Estimation of holding time, where the error bars represent the range of duplicate results.

Summary

This chapter has discussed the important topic of sampling. The complete chain of events, from obtaining a laboratory sample from bulk material, through to storage and subsampling within the laboratory, has been considered. This chapter has identified the different types of samples that may be required and the sampling plans that are used to obtain them. Procedures for subsampling, storage and stabilization of samples are also described. An understanding of these topics is important as even if the analytical method is fully validated and used correctly, the results obtained will be of little use if the sample measured has not been obtained, stored and handled correctly.

References

1. Horowitz, W., *Pure Appl. Chem.*, **62**, 1193–1208 (1990).
2. de Zorzi, P., Barbizzi, S., Belli, M., Ciceri, G., Fajgelj, A., Moore, D., Sansone, U. and van der Perk, M., *Pure Appl. Chem.*, **77**, 827–841 (2005).
3. 'Draft General Guidelines on Sampling', Document reference ALINORM 04/27/23, Appendix III, Codex Alimentarius Commission, 2004. [http://www.codexalimentarius.net] (accessed 7 November, 2007).
4. 'Water Quality – Sampling – Part 1: Guidance on the Design of Sampling Programmes and Sampling Techniques', ISO 5667-1:2006, International Organization for Standardization (ISO), Geneva, Switzerland, 2006.
5. 'Sampling Procedures for Inspection by Attributes – Part 10: Introduction to the ISO 2859 Series of Standards for Sampling for Inspection by Attributes', ISO 2859-10:2006, International Organization for Standardization (ISO), Geneva, Switzerland, 2006.

6. 'Sampling Procedures for Inspection by Attributes – Part 1: Sampling Schemes Indexed by Acceptance Quality Limit (AQL) for Lot-by-Lot Inspection', ISO 2859-1:1999, International Organization for Standardization (ISO), Geneva, Switzerland, 1999.
7. 'Sampling Procedures for Inspection by Attributes – Part 2: Sampling Plans Indexed by Limiting Quality (LQ) for Isolated Lot Inspection', ISO 2859-2:1985, International Organization for Standardization (ISO), Geneva, Switzerland, 1985.
8. 'Sampling Procedures for Inspection by Attributes – Part 3: Skip-Lot Sampling Procedures', ISO 2859-3:2005, International Organization for Standardization (ISO), Geneva, Switzerland, 2005.
9. 'Sampling Procedures for Inspection by Attributes – Part 4: Procedures for Assessment of Declared Quality Levels', ISO 2859-4:2002, International Organization for Standardization (ISO), Geneva, Switzerland, 2002.
10. 'Sampling Procedures for Inspection by Attributes – Part 5: System of Sequential Sampling Plans Indexed by Acceptance Quality Limit (AQL) for Lot-by-Lot Inspection', ISO 2859-5:2005, International Organization for Standardization (ISO), Geneva, Switzerland, 2005.
11. 'Sampling Procedures for Inspection by Variables – Part 1: Specification for Single Sampling Plans Indexed by Acceptance Quality Limit (AQL) for Lot-by-Lot Inspection for a Single Quality Characteristic and a Single AQL', ISO 3951-1:2005, International Organization for Standardization (ISO), Geneva, Switzerland, 2005.
12. 'Codex Standard for Food Grade Salt', CODEX STAN 150-1985, Amend. 3, 2006. Codex. [http://www.codexalimentarius.net] (accessed 7 Novemeber, 2007).
13. Codex Alimentarius MRLs, Codex. [http://www.codexalimentarius.net] (accessed 7 November, 2007).
14. 'Portion of Commodities to which MRLs Apply', Codex Alimentarius Volume 2A, Part 1 – 2000, Section 2.1, 2000, Codex. [www.codexalimentarius.net] (accessed 7 November, 2007).

Chapter 4

Preparing for Analysis

Learning Objectives

- To be able to identify the factors which have to be considered when choosing a method of analysis.
- To know where to look for suitable methods.
- To understand and be able to identify the causes of unsatisfactory results.
- To understand how to validate analytical methods.

4.1 Selecting the Method

As discussed in Chapter 1, analysis involves the determination of the composition of a material, i.e. the identification of its constituent parts and in many cases how much of each is present and, sometimes, in what form each is present. This chapter describes the process of selecting a suitable analytical method to carry out such determinations and how to check that the procedure selected is adequate for the job in hand. Before starting work on a sample, it is vital to enquire why the analysis is being done and what will happen to the result(s) and what decisions will be taken based on the constituent parts identified and the numerical values obtained. It is essential that the requirements of the customer, internal or external, are fully understood. Which property needs to be measured (the measurand)? For example, is it the total amount of iron in a tablet that is required or the amount extracted into stomach acid simulant?

Quality Assurance in Analytical Chemistry E. Prichard and V. Barwick

DQ 4.1

The purpose of an analysis and the use to which the analytical report or certificate of analysis might be put are numerous. Can you suggest some of these?

Answer

Some examples were given in Chapter 1 but hopefully you will have thought of additional ones. You might have suggested any one, or all, of the following:

- Preparation of a data bank of figures for nutritional content of a typical food basket;
- Release of a batch of product for sale;
- Analysis of a product to determine its tariff classification;
- Prosecution of a company for selling an incorrectly labelled product; e.g. a sausage containing insufficient meat, or containing pork instead of beef;
- Prosecution of a driver for being over the drink/drive limit based on the analysis of a blood sample.

Chapter 1 dealt with this in some detail. What we now need to know are the criteria to consider in selecting a method fit for our purpose and where such methods may be found.

4.2 Sources of Methods

Selection of a suitable analytical method can be made once the reason for carrying out the analysis is well understood. Analytical methods may be (a) qualitative or (b) quantitative or semi-quantitative. The former usually pose few problems if only an indication is required as to whether a particular analyte is present or not – certainly not how much with a value having a small uncertainty. If a negative result is required (i.e. confirmation of absence from the product), then one has only to worry about the limit of detection of the test used. Many tests to confirm the absence of impurities in pharmaceutical products fall into this category. Equally, rapid tests for positive confirmation are often made on unknown substances. These may subsequently be confirmed by other, quantitative tests. Quantitative methods are used in a variety of situations and a variety of different methods can be employed. What you must always remember is that the method used must be fit for the purpose.

Suitable methods fall into a number of categories and there are many sources where methods may be found, as follows:

- In-house methods developed by one laboratory for their own special needs.
- Methods published in the open scientific literature, e.g. *The Analyst*, *Journal of AOAC International*, *Journal of the Association of Public Analysts*, *Journal of Chromatography*, etc.
- Methods supplied by trade organizations, e.g. the Institute of Petroleum (Energy Institute).
- Methods in books published by professional organizations, e.g. The Royal Society of Chemistry (Analytical Methods Committee), Association of Official Analytical Chemists, etc.
- Methods from standards organizations, e.g. (UK) BSI, BP; (International) ISO; (Europe) CEN, EP; (USA) ASTM, EPA, USP, etc.
- Methods from statutory publications, e.g. The Fertilisers (Sampling and Analysis) Regulations 1991 (SI No. 973) HMSO.

The degree of validation of the methods may be quite different. What validation means is that the method has been subject to a study which shows that, as applied in the user's laboratory, it provides results which are fit for their intended purpose. The method satisfies some pre-defined criteria. When standard or internationally agreed methods are being developed, the validation of the method is more complicated and time-consuming than that of methods developed in-house. Such validation involves a collaborative study using analysts working in a number of laboratories. This has already been mentioned in Chapter 1 and the organization of collaborative studies is discussed in Chapter 7. However, this more elaborate procedure does not necessarily mean that the method is more reliable than in-house methods.

In the field of trace analysis where analysts are attempting to determine very low levels of analytes ($mg\,kg^{-1}$, $\mu g\ kg^{-1}$) in a very complex matrix, e.g. food or agricultural products, it is often necessary to examine large numbers of samples using methods that might take anything from a few minutes to a whole week to complete. The very rapid methods can be used to eliminate the majority of samples containing no detectable analyte so that more expensive resources can be devoted to those samples where there is evidence for a positive result, (presumptive-positive). The quick methods are generally less reliable than those taking a whole week for which very expensive analytical instrumentation, e.g. a mass spectrometer, is required. Hence, in this type of work it is often convenient to divide methods into different categories. Methods can be categorized in a number of ways. It could be on the basis of whether one is expecting to find some of the analyte in most samples, (surveillance) or none is expected, (screening).

Table 4.1 Classification of methods by degree of confidence

Method	Increasing degree of confidence (↓)
Screening	
Surveillance	
Accepted	
Standard	
Regulatory	
Reference	
Primary	

Another way is by the confidence required by the customer or other users of the results. This applies to both qualitative and quantitative methods. It hinges on the consequences of making a wrong decision. In some cases, the consequences may be modest and then the degree of confidence in each result can be lower than where there are serious consequences if wrong decisions are made. Some of the terms used to indicate the degree of confidence offered by an analytical method are listed in Table 4.1.

There are two main types of routine analysis carried out, i.e. those required for screening purposes, where one is testing a large number of individual unrelated samples, e.g. work-place drugs screening, and those for surveillance activities, e.g. monitoring foodstuffs for the level of toxic metals.

Screening methods must be extremely rapid, permitting a high throughput of samples at low cost. A small number of false positive results (i.e. where the analyte is detected but is not actually present) is acceptable since these will be eliminated by further studies. The method should be sufficiently sensitive to eliminate false negatives (i.e. where an analyte which is present is not detected). These methods can be qualitative or semi-quantitative, and may be validated only to the extent of the limit of detection, by the operational laboratory. Surveillance methods are very similar to screening methods but are usually somewhat less rapid, with a lower throughput of samples although they do yield quantitative results. These may be developed and validated in-house and the judgement of suitability will be made by the laboratory. The selectivity will be better than for screening methods but may not be unambiguous. Further measurements may be necessary.

Accepted methods usually represent a consensus view from a number of analytical laboratories working in a particular application area. They may be developed and validated collectively under the auspices of professional or official bodies or trade organizations. Standard methods are similar to accepted methods but are usually developed on a national or international basis by an organization with some official status. These methods are usually published and have detailed procedures. The extent of validation of such methods can be taken for granted but has to be examined carefully.

Regulatory and reference methods are typically identified by an official body for use in enforcement of a specific regulation. The methods can be of two types, i.e. confirmatory or reference. The former is used following a presumptive positive identification obtained using a screening method and will involve a detection system based on a different physico-chemical principle. Alternatively, co-chromatography (see Section 4.4.1) can be used to confirm identity. Although these methods will have been fully validated, the laboratory using the method has to verify that it can achieve the published performance requirements. Reference methods will also have been fully validated and tested by an approved collaborative study in which satisfactory performance data covering bias and precision have been obtained. Detailed procedures for its use to produce data within the claimed limits will be available. These types of method are frequently used to characterize reference materials.

Primary methods have the highest metrological qualities, whose operation can be completely described and understood, and for which a complete uncertainty statement can be written down in terms of SI units. Such methods are used by national laboratories participating in the development of a national or international chemical measurement system.

In brief, the factors that have to be considered when choosing between types of method include the following:

- type of sample, matrix and measurand;
- time;
- cost;
- equipment availability;
- frequency of false negatives and false positives;
- limit of detection;
- limit of quantitation;
- working range;
- selectivity;
- quantitative or semi-quantitative;
- performance requirements, including extent of validation necessary;
- acceptable measurement uncertainty.

4.3 Factors to Consider when Selecting a Method

Having established the purpose of an analysis, a decision needs to be made about which particular analytical procedure to use.

DQ 4.2

What factors will influence your choice of method? List the factors that could be used to distinguish one method from another.

Answer

The choice of method depends on the purpose for which the analysis is being performed. The customer requesting the analysis may specify the method to be used. Even in this situation, it is the responsibility of the laboratory to demonstrate that the method is capable of producing results that are reliable. When no method is specified the points to consider have already been identified in Section 4.2. The acceptable level of measurement uncertainty specified or implied will, to a certain extent, set the precision and bias levels. All of the topics covered in the following sections may be crucial, depending on the purpose of the analysis, and should appear on your list.

4.3.1 Limit of Detection

The limit of detection of an individual analytical procedure is the smallest amount of an analyte in a sample which can be detected but not necessarily quantified with an acceptable uncertainty. The limit of detection is derived from the smallest concentration, x_L that can be detected with reasonable certainty for a given procedure. At this concentration, the method will indicate that there is some measurand present, at the stated level of significance, but the amount cannot be specified. The value x_L is given by equation (4.1):

$$x_L = \overline{x}_{bl} + k s_{bl} \qquad (4.1)$$

where $\overline{x}_{bl}$ is the mean of the results obtained by measuring blank solutions, s_{bl} is the standard deviation of the blank measures and k is a numerical factor chosen according to the confidence level required.

For many purposes, only an approximate value of the limit of detection is required and this is calculated by using either equation (4.2a) or (4.2b). If the instrument signal-to-noise ratio is obtained in terms of the response it will need to be converted to concentration units:

$$\text{LoD} = 3s_{bl} \qquad (4.2a)$$

$$\text{LoD} = 3 \times \text{signal-to-noise ratio} \qquad (4.2b)$$

This approximation is probably adequate during method validation as it provides an indication of the concentration below which detection becomes problematic. It indicates that a signal more than $3s_{bl}$ above the sample blank value could have arisen just from the blank in fewer than 5% of the measurements and therefore there is a 95% probability that it has arisen from the measurand. Where the work is to support regulatory or specification compliance, a more exact approach may be required [1, 2]. When a blank solution cannot be used, it can be replaced with a solution containing a low level of measurand. The limit of detection (LoD) is especially important in trace analysis, when one has to decide whether a contaminant is present below or above the legal limit. Ideally, the limit of detection of the method selected should be at least one-tenth of the concentration set as the legal limit. For example, if the legal limit for lead in tap water is 50.0 $\mu g\ l^{-1}$ the analytical method used should be capable of measurements down to 5.0 $\mu g\ l^{-1}$.

In some cases, a limit of quantitation (quantification) may need to be considered where it is necessary not only to detect the presence of an analyte but also to determine the amount present with a reasonable statistical certainty. The limit of quantitation of an individual analytical procedure is the smallest amount of an analyte in a sample, which can be quantitatively determined with acceptable uncertainty. More detail can be found in Section 4.6.4.

4.3.2 Precision

Precision is the closeness of agreement between independent test results obtained under stipulated conditions. Precision depends only on the distribution of random errors and does not relate to the true value. It is calculated by determining the standard deviation of the test results from repeat measurements. In numerical terms, a ***large*** number for the precision indicates that the results are scattered, i.e. the precision is poor. Quantitative measures of precision depend critically on the stipulated conditions. *Repeatability* and *reproducibility* are the two extreme conditions.

Repeatability (r) is the value below which the absolute difference between two single test results obtained with the same method on identical test material, under the **same conditions** (same operator, same apparatus, same laboratory and a short interval of time) may be expected to lie, with a specified probability; in the absence of other indications, a probability of 95% is used.

Reproducibility (R) is the value below which the absolute difference between two single tests results obtained by the same method on identical test material, under **different conditions** (different operators, different apparatus, different laboratories and/or different time) may be expected to lie, with a specified probability; in the absence of other indications, the probability is again taken as 95%.

The precision limits r and R are given by equations (4.3a) and (4.3b), respectively, where $t_{\nu,\alpha}$ is the Student t-value for ν degrees of freedom and α corresponds to the stated probability, s_r is the repeatability standard deviation and s_R is the reproducibility standard deviation calculated from $(\nu + 1)$ results:

$$r = t_{\nu,\alpha}\sqrt{2}s_r \tag{4.3a}$$

$$R = t_{\nu,\alpha}\sqrt{2}s_R \tag{4.3b}$$

If repeatability is the only estimate of precision that is obtained, this is unlikely to be representative of the variability observed when the method is used over a long period of time. **Intermediate precision** is often more relevant – this expresses the within-laboratory variation or within-laboratory reproducibility (different days, different analysts, different equipment, etc.). This is initially obtained from validation studies and confirmed later by examining the results obtained for quality control material measured over a period of about three months (see the quality control (QC) charts in Chapter 6).

High precision is not always required. If, for example, you are merely trying to establish whether the fat content of biscuits falls within the range 20–30%, a high degree of precision may not be necessary unless the result obtained lies close to the margins. More information can be found in Section 4.6.2.

4.3.3 Bias/Recovery

Measurements are subject to systematic errors as well as the random errors covered in Section 4.3.2. *Bias* is the difference between the mean value of a large number of test results and an accepted reference value for the test material. The bias is a measure of 'trueness' of the method. It can be expressed in a number of ways, i.e. simply as a difference or as a ratio of the observed value to the accepted value. This latter representation, when expressed as a percentage, is often termed *recovery*. This represents how much of the analyte of interest has been extracted from the matrix and measured. This is dealt with in Section 4.6.3.

4.3.4 Accuracy

Accuracy is often used to describe the overall doubt about a measurement result. It is made up of contributions from both bias and precision. There are a number of definitions in the Standards dealing with quality of measurements [3–5]. They are only different in the detail. The definition of accuracy in ISO 5725-1:1994, is 'The closeness of agreement between **a test** result and the accepted reference value'. This means it is only appropriate to use this term when discussing a single result. The term 'accuracy', when applied to a set of observed values, describes the consequence of a combination of random variations and a common systematic error or bias component. It is preferable to express the 'quality' of a result as its *uncertainty*, which is an estimate of the range of values within which, with a specified degree of confidence, the true value is estimated to lie. For example, the concentration of cadmium in river water is quoted as 83.2 ± 2.2 nmol l^{-1}; this indicates the interval 'bracketing' the best estimate of the true value. Measurement uncertainty is discussed in detail in Chapter 6.

Very often a high degree of accuracy, i.e. a small number after the $\pm$ in the example above, is not important. This might be the case for trace analysis where the concentration of the contaminant is well below the permitted level. For example, the permitted maximum residue level of fluorine in complete animal feeding stuffs for pigs is $100\,mg\,kg^{-1}$. If a sample is analysed and found to contain $30\,mg\,kg^{-1}$, it does not matter if the analysis is in error by as much as 100%, as the measured level of contamination is still well below the permitted maximum. Where the concentration of a contaminant, or permitted additive, is close to the maximum amount allowed, accuracy becomes more important (see Chapter 6, Figure 6.15). As an extreme example, a determination of the amount of gold in a bullion bar will always demand a very high degree of accuracy (to within 99.99% of the true value) if large sums of money are not to be lost (or gained).

4.3.5 Time

If a large number of samples have to be analysed, a method that is simple and rapid is to be preferred so that data can be acquired quickly and with the minimum of effort and cost. As a result of this initial survey, you might be able to decide between:

- There is no problem and, therefore, no further work is required;

and

- There is some evidence that a particular analyte may be present but this needs to be confirmed by further measurements. This might involve the use of the same method on additional samples, or an alternative method which takes longer to carry out but can produce results that have a lower uncertainty.

4.3.6 Equipment Required

All items of equipment must be considered, including balances and volumetric measuring devices, not just the expensive equipment. In terms of instrumentation, while a method using a mass spectrometer may be ideal for the study, if no such equipment is available the job will have to be contracted out to another laboratory, or another approach agreed with the customer. Neutron activation or radiochemical measurements require special equipment and dedicated laboratory facilities and safety procedures. Such techniques are often not generally available and are better left to specialist laboratories.

4.3.7 Sample Size

In many industrial areas, as well as food and agriculture, the amount of sample available to the analyst is not normally a limiting factor. However, in clinical chemistry the opposite applies, as no patient is willing to donate large volumes of blood for analysis! Similarly in forensic work, the sample material may also be limited in size. Sample size is linked to the limit of detection. Improved detection levels can sometimes be achieved by taking a larger mass of sample. However, there are limits to this approach. For example, where organic matter has to be destroyed using oxidizing acids, the smaller the mass of sample taken the better, as the digestion takes less time and uses smaller volumes of acids, thus giving lower blank values. Where a large mass of sample is essential, destruction of organic matter is preferably carried out by dry-ashing in a muffle furnace. Where the sample to be analysed is not homogeneous, use of a small test sample should be avoided because the portion of sample used in the analysis may not be truly representative of the bulk material and could give rise to erroneous results (see 'Sampling' in Chapter 3).

4.3.8 Cost

Most analytical chemists and their customers have to be concerned with the cost of an analysis. While the major factors are the human resource and the cost of running and maintaining a laboratory, the choice of method may have a small bearing on the total cost of the job. Analysis of a single sample will always be charged at a higher rate pro-rata than a batch of six. Analysis requiring techniques such as mass spectrometry or nuclear magnetic resonance spectroscopy will be more expensive than classical techniques because of the capital cost of the equipment used and the seniority of the staff required to interpret the data produced by such techniques. A customer may be prepared to accept the risk of making decisions based on results with a large known uncertainty rather than incur the extra cost of obtaining results that have a smaller uncertainty.

4.3.9 Safety

The need for special facilities for work involving neutron activation analysis and radiochemical measurements has been referred to above in Section 4.3.6. Other safety factors may also influence your choice of method. For example, you may wish to avoid the use of methods which require toxic solvents, such as benzene and certain chlorinated hydrocarbons, or toxic reagents, such as potassium cyanide, if alternative procedures are available. Where Statutory Methods have to be used, there may be no alternative. In such cases, it is essential that staff are fully aware of the hazards involved and are properly supervised. Whatever method is used, the appropriate safety assessment must be carried out before the work is started. Procedures should be in place to ensure that the required safety protocols are followed and that everyone is aware of legislative requirements.

4.3.10 Selectivity

Selectivity refers to the extent to which a method can be used to determine particular analytes in mixtures or matrices without interference from other components of similar behaviour [6]. The analytical requirement will have been established (see Section 4.1) so that the measurand is well-defined. The degree of discrimination between the measurand and other substances present in, or extracted from, the matrix must be carefully considered. It is necessary to be sure that the identity of the measurand is unequivocal. Attention will have to be paid to the clean-up procedures used and the discriminating power of the detection system. It may be necessary to carry out tests where potential interferents, that may be present in some samples, are added to the matrix and their influence measured. This is part of method validation (see Section 4.6).

4.3.11 Making Your Choice

Ultimately, the choice of method will depend on several factors. Above all, 'fitness for purpose' must be uppermost in your mind. Will the method you have selected be adequate for the decision you and/or your customer has to take when the result is available?

Once you have a clear picture in your mind as to why the analysis is being carried out and what you hope to achieve, carry out a literature survey and identify one or more methods/procedures that appear to satisfy the criteria set. Frequently, more than one technique can be used to detect the same analyte.

DQ 4.3

What techniques are available for the determination of trace metals?

Answer

Your list should have included the following:

- colorimetry;
- atomic absorption spectrometry (flame and furnace);
- inductively coupled plasma (ICP)–atomic emission spectrometry.

You may also have included techniques, such as:

- anodic stripping voltammetry;
- ion chromatography;
- ICP–mass spectrometry;
- X-ray fluorescence;
- neutron activation analysis;

and possibly others.

It is important to remember at this stage that whatever is the chemical entity to be determined, there are usually several techniques that can be used for the measurement. Your problem is to select the best approach for the job in hand. Some techniques can be quickly eliminated because the equipment is not available. However, several options may still remain.

SAQ 4.1

The concentration of copper in a sample may be determined by using an iodometric titration or by atomic absorption spectrometry. In each of the following examples, calculate the cost of the assay (assume that the charge for the analyst's time is £50 per hour):

(a) The determination of copper in a copper sulfate ore by reaction with KI, and iodometric titration.

(b) The determination of low levels of copper in a pig feed by wet-digestion and atomic absorption spectrometry. In this example, it is possible to carry out the digestion and extraction step on two test portions at the same time.

It is now necessary to discuss in more detail the performance criteria one can use to evaluate different methods and to describe the validation of different analytical procedures so that you can decide whether or not a given method will fulfil your own particular requirements. In many cases, there will be no method which is entirely suitable for your purpose. In such cases, it will be necessary to adapt an existing method. Before use, such an amended method will need to be validated to ensure that the modifications introduced do not produce erroneous results (see Section 4.6).

4.4 Performance Criteria for Methods Used

In Section 4.3, some factors which need to be considered in choosing a method of analysis were discussed in general terms. The next step is to consider the properties of a method that will enable a choice to be made. This is done for a specific case, e.g. the determination of residues of chemicals used in veterinary practice to treat animal diseases and to prevent the development and spread of disease where large numbers of animals are kept in close proximity to each other. Such chemicals may be administered by injection, or orally as a constituent of the feed. Some chemicals are metabolized and excreted while others may be partially retained in edible products such as milk, eggs, meat and offal (liver or kidney). The detection and determination of such residues is a very difficult analytical problem.

DQ 4.4

Suggest reasons why the determination of veterinary residues in animal products presents difficulties for the analyst.

Answer

Your answer should have included the following:

- the level of residues present is likely to be very low, in the region of $\mu g\ kg^{-1}$, and therefore a method capable of detecting very low levels of the compound is required;
- a number of different compounds are in use and in many cases the analyst will not know which product has been administered;
- there may be a problem in getting a representative sample;
- the samples may require extensive pre-treatment to get the analyte in a form suitable for determination;
- some chemicals occur in tissues in a form that is different from that administered – they may have been metabolized (e.g. hydrolysed, oxidized) or bound to tissue constituents.

Congratulations if you thought about the last one and if you realized that the analysis is made more difficult by the presence of large numbers of co-extracted compounds. Hence, a method involving extensive 'clean-up' or purification of the initial extract will be required. In addition, the detection system selected will not only need to be able to measure very low concentrations but also be highly selective to ensure that positive signals are not obtained from co-extracted analytes.

Since the analytical problem is so difficult, there will not be many methods or techniques available which are satisfactory for the purpose required, i.e. to determine whether residues are present at or above the legal limits. There will be instances where no method is available which matches the criteria initially specified for precision and accuracy. In these cases, the customer must be fully informed of the situation, since developing a new method takes time and money.

There is some help in terms of setting some of the performance characteristics. The European Community implementing Council Directive 96/23/EC has considered the level of bias (trueness) and precision appropriate for analytical methods used to monitor the concentrations of certain substances and residues of the substances in animal products for concentrations ranging from $1\ \mu g\ kg^{-1}$ to $1\ mg\ kg^{-1}$. Their recommendations for the trueness and precision of analytical methods are shown in Tables 4.2 and 4.3, respectively. However,

Table 4.2 Minimum trueness of quantitative methods

True content (μg kg^{-1} (mass fraction))	Acceptable range (%)
≤ 1	−50 to 20
1 to 10	−30 to 10
≥ 10	−20 to 10

Table 4.3 Relationship between precision (reproducibility) and concentration level

Content (μg kg^{-1} (mass fraction))	*CV* (%)
1	(45.3)[a]
10	(32)[a]
100	23
1000[b]	16

[a] For concentrations lower than 100 μg kg^{-1}, equation (4.4) gives values of %*CV* that are unacceptably high. The %*CV* value should be as low as possible. This equation becomes less helpful for measurements at very low concentrations, e.g. on veterinary residues.
[b] 1 mg kg^{-1}.

residues are unlikely to be found in the higher ranges quoted. The interlaboratory percentage coefficient of variation (%*CV*) given for the repeated analysis of a reference material under reproducibility conditions is that calculated by the Horwitz equation (equation (4.4)) [7]. The Horwitz function is shown later in Figure 4.6 and discussed with some refinements in Section 4.6.2:

$$\%CV = 2^{(1-0.5\log C)} \text{ or } s_R = 0.02C^{0.8495} \tag{4.4}$$

where s_R is the inter-laboratory reproducibility standard deviation and C is the mass fraction (g g^{-1}) expressed as a power of 10 (e.g. 1 mg kg^{-1} = 10^{-6}, log $C = -6$) and:

$$\%CV = 100 \times \frac{s_R}{C} \tag{4.5}$$

The values quoted in Table 4.3 refer to the spread of results expected when a given sample is analysed in a number of separate laboratories. For repeat analyses carried out by one operator in a single laboratory, the coefficient of variation (%*CV*) would typically be one half to two thirds of the values shown in Table 4.3. For within-laboratory reproducibility (intermediate precision), the %*CV* should not be greater than the reproducibility %*CV* for the given concentration in Table 4.3.

Note that it is more difficult to analyse low concentrations.

DQ 4.5

Why is the spread of the results from a number of laboratories likely to be larger than the spread of results obtained from one laboratory?

Answer

Determinations made in several laboratories are likely to show a large degree of variation since different batches of reagents (and from different suppliers) will have been used. In many cases, different equipment will have been used, the analysts will vary in competence, they may have received different training and their experience will differ. The environmental factors could have an effect on the results, e.g. temperature changes, contamination from other work in progress and possibly lighting effects. In a single laboratory, these factors are likely to vary less, thus reducing the variation in the results obtained.

You may also find it useful to read Section 6.3 in Chapter 6 ('Measurement Uncertainty') at this stage.

Equally, one expects to obtain more accurate results (closer to the true value) at higher concentrations of analyte. While −50% to +20% given in Table 4.2 may seem an unacceptably large range, it is based partly on what can be achieved in practice. Furthermore, one must remember that even legal limits are quoted with large uncertainty limits because the values quoted depend on toxicological assessments. The analyst is still far in advance of the toxicologist as far as accuracy and precision of measurements are concerned!

DQ 4.6

How do these values for precision and bias fit in with your own requirements?

Answer

The appropriate levels will depend on what is being measured and the reason for carrying out the measurement. This is always an interesting discussion topic.

There is a saying that 'the strength of a chain is no greater than the strength of its weakest link'. In analytical chemistry, this means that all parts of a method are vital to the success of the determination. Nevertheless, much depends on the limit of detection and selectivity of the detection system used at the final stage of the method. This is why, in many cases, the method of detection is selected first.

Then, the extraction and 'clean-up' stages can be tailored to meet the requirements of the particular detector being used. We can now consider some techniques used at the final separation and detection stages and discuss criteria which enable a decision to be made about which technique is the most appropriate. So often, experts in a particular technique believe that their technique can solve all of the world's problems. It may well be able to detect the analyte in question, but is it the most suitable method of detecting that analyte in the given matrix?

4.4.1 Criteria for the Determination of Analytes by Selected Techniques

4.4.1.1 Thin Layer Chromatography

Quite elegant separations can be achieved by using thin layer chromatography (TLC), particularly when using two-dimensional chromatography. Detection systems range from the visual identification of coloured compounds to spraying with reagents to form a coloured derivative on the plate. Some compounds fluoresce under UV irradiation. Special cabinets are available for use in such cases. If this technique is going to be used for pesticide analysis, then two-dimensional high performance TLC (2D HPTLC) with 'co-chromatography' is required.

In all TLC work, identification is confirmed by measurement of the distance travelled along the plate by the analyte compared to the solvent front, the R_f value, and by reference to standard solutions run on the same plate. Further confirmation can be obtained by using a reference sample and measuring this under the same conditions as the sample. This allows measurement of the R_x value. This is illustrated in Figure 4.1.

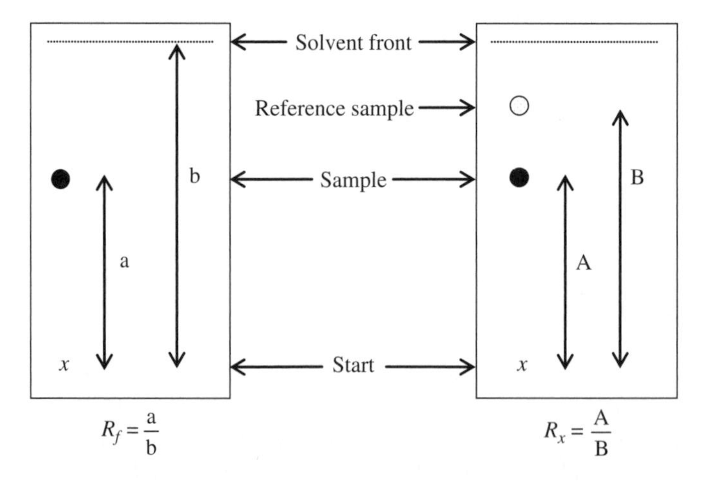

Figure 4.1 Schematic diagram for determining R_f and R_x values.

The definitions of R_f and R_x are as follows:

$$R_f = \frac{\text{distance travelled by the analyte}}{\text{distance travelled by the solvent front}} \tag{4.6}$$

and:

$$R_x = \frac{\text{distance travelled by the analyte}}{\text{distance travelled by a reference sample}} \tag{4.7}$$

The visual appearance of the spot produced by the sample extract should be indistinguishable from that produced by the reference sample of the analyte in both size and shape. Spots produced by other co-extractives should be separated from the analyte spot by a distance equal to half the sum of the spot diameters. The R_f value of the spot produced by the extract should be within $\pm$ 5% of the R_f value obtained with the reference sample of the analyte. Further confirmation can be achieved by 'co-chromatography', i.e. 'over-spotting' a sample extract with reference analyte solution and developing the chromatogram. No additional spot should be obtained. In two-dimensional chromatography, the R_f value should be checked in both directions. Alternatively, the spot may be cut out, the analyte eluted and then examined further by spectroscopic or other techniques. The absorption spectrum of the sample should not be visually different from the full absorption spectrum of the standard. If computer-aided library searching and matching is used, then a critical level should be set for the 'match factor'. For quantitative measurements, standard solutions close in concentration to that present in the sample extract should be added to the plate. Alternatively, densitometric equipment can be used if the spots are regular in shape. A peak is obtained for each spot and the peak can be evaluated using height or area measurements by comparison with standard spots on the same plate.

It is always worth trying another solvent to check if you can get any discrimination or separation.

4.4.1.2 Gas and Liquid Chromatographic Separation

Either gas chromatography (GC) or liquid chromatography (LC) can be used as a separation technique coupled with a variety of detection methods. Mass spectrometry (MS) is one of the most popular means of detection. When using GC–MS, a capillary column should be used, while any suitable LC column can be used for LC–MS. It is advisable to obtain a print-out of the chromatogram so that the shapes of individual peaks can be assessed. Electronically produced data using integrators should be treated with some suspicion and always examined visually to check the selected baseline, start- and end-points of peak integration, etc.

Chromatographic conditions should be optimized wherever possible to achieve baseline separation of the analyte peak from other peaks produced by co-extracted compounds. The retention time of the analyte should be at least twice the column void volume and the retention time of the analyte peak in the sample extract should agree with that of the pure analyte peak within a margin of $\pm$ 0.5%. Internal standards of reference material should be used whenever possible. Added at the injection stage, they serve to check the volume of extract added to the column. This is particularly important in GC where very small sample volumes are used and may be injected manually. If the internal standard is added to the sample before extraction it serves to validate the recovery throughout the entire procedure (see Section 4.6.). Columns of different polarity can be used as a check on the identity and purity of the analyte since different retention times will be obtained. The internal standard should be structurally related to the analyte. If this is not possible, co-chromatography should be used. This involves adding a known amount of the analyte to the test portion and this should result in no additional peaks. The peak increases in size by the amount added (taking into account any dilution effects). The retention time should not change by more than 5% and the peak width at half-peak-height should not change by more than 10%.

The MS detection system can be such that the full mass spectrum is observed (at least five peaks) or just selected ions monitored (SIM) with three or four identification points. For some analyses, it may be necessary to use MS–MS^n techniques [8]. In LC–MS, it is important to make sure that ionization of the compounds of interest has been achieved. For all of these approaches, the criteria for matching of the analyte with the standard should be established during validation studies.

Liquid and gas chromatography can also be used with other detection techniques. When LC is used with UV/VIS detection, the shape of the analyte peak should be examined carefully for the presence of co-eluting interferences. This is done by checking the retention time. The latter must agree with that obtained using pure analyte solution. The purity of the analyte peak can be confirmed by using co-chromatography or by using a diode-array detector. The wavelength of maximum absorption of the peak produced by the sample extract must be equal to that produced by pure analyte, within the resolving power of the detector. The spectra produced by (a) the leading edge (b) the apex and (c) the trailing edge of the peak from the sample extract must not be visually different from each other, or from that produced by the pure analyte. The purity of the peak can be checked automatically by using a diode-array detector. Detection using a single wavelength gives far less information.

Gas chromatography can also be used with other detection systems, e.g. electron-capture detection (GC–ECD) and flame-ionization detection (GC–FID). The optimization will have been carried out during validation studies. The peak separation criterion is the same as that given for LC above. Once again, co-chromatography can be used for confirmation.

There are other techniques that could be used but it is not the purpose of this book to describe them. This section is intended to illustrate the point that often one has to select a separation technique and then an appropriate detection system.

It is clear that both need to be 'fit for purpose'. In spite of care in selecting the techniques, things can still go wrong and incorrect results produced.

4.5 Reasons for Incorrect Analytical Results

Before considering how one can ensure that analytical data obtained are correct and fit for the purpose required, it is worth thinking about what ***could*** go wrong. Then it will be easier to work out how to avoid making mistakes.

DQ 4.7

Why do you think analytical results are sometimes wrong?

Answer

You may well have come up with a list that looks like this:

- incompetence;
- calculation/transcription errors;
- unsuitable method used;
- contamination;
- interferences;
- calibration errors;
- sampling errors;
- losses/degradation.

Let us now look at these in a little more detail.

4.5.1 Incompetence

People who organize collaborative studies will tell you that there is always someone who sends in results that are widely different from those produced by other collaborators – by from one to several orders of magnitude! In many cases, this is caused by mistakes in calculation, or forgetting a dilution factor, etc. In some cases, it may be that results are calculated in the wrong units or insufficient care is taken with the units, e.g. mixing mg and g or not taking the correct conversion factor. Poor laboratory practice, e.g. storing 200 ml and 250 ml graduated flasks close together can also lead to mistakes, in this case 50 in 200, or 25%! Errors in the labelling of samples and equipment used in subsequent analysis can also occur. Spectrophotometric measurements on solutions that are not optically clear will be falsely high. Maybe it has been assumed that staff are competent in the analytical technique because they have received training but their competency has not been checked. There are probably many other human errors resulting in incorrect results. These may include the use of unchecked spreadsheets, calculation errors and/or transcription errors which pass unnoticed or unchecked.

4.5.2 Method Used

Erroneous results may be obtained even with approved methods if they are used outside of the tested calibration range or with matrices that were not included in the original validation process. For example, the presence of fat often causes problems in trace organic analysis. Can the method used cope with the fat actually present in the sample? Is the digestion technique appropriate for that particular sample matrix? If a method validated for analysing water with negligible organic matter is used to analyse water with a high humic content, the results may be inaccurate.

Many people take an approved method and introduce subtle changes in the procedure to suit their own circumstances or convenience. These may include changes to the sample weight/reagent ratios, times and temperatures used which may be critical and so invalidate the method. Changes to the recommended purity of reagents used, using reagents that have passed their expiry date, or by changing the reagent supplier, can all influence the results obtained. The moisture content of samples, reagents and alumina used in adsorption chromatography are further examples where care is required. The extent to which a method may be modified, without reducing the quality of the result, is what is meant by the *ruggedness* or *robustness* of a method.

4.5.3 Contamination

It is vital to know if the analyte is present in the laboratory environment, adsorbed on the glassware, in reagents or in the demineralized water used in the analysis. These are all potential sources of contamination. This is particularly important when carrying out new determinations and for trace analysis. It is also important to ensure that your colleagues working nearby are not using chemicals that could affect your determination.

4.5.4 Interferences

In addition to the analyte, the matrix will contain many other compounds. The method chosen must discriminate between the analyte of interest and other compounds also present in the sample. The test portion may have to pass through many analytical stages before the analyte is obtained in a form suitable for final measurement. First, the analyte may need to be separated from the bulk of the sample matrix. Further treatment may then be required to obtain an aliquot that is sufficiently 'clean' (i.e. free from potential interferences) for the end-measurement technique. A general scheme of analysis is presented in Table 4.4 to illustrate the different approaches used depending on the nature of the analyte and of the matrix.

Table 4.4 General scheme for determination of measurands

Matrix	Analyte	
	Inorganic	Organic
Inorganic	(a) Separate analyte from other inorganic analytes using classical methods, ion-exchange chromatography or complexation reactions (b) Use a specific detection system	(a) Separate analyte from matrix using solvent extraction (b) Determine analyte by, e.g. GC–MS
Organic	(a) Destroy matrix by oxidation (dry-ash or inorganic acids) (b) Separate and determine by using a specific detection system	(a) Separate analyte from matrix, e.g. by solvent extraction or solid-phase extraction (b) Separate analyte from co-extracted compounds by distillation, partition, chromatography, etc. (c) Concentrate measurand (if necessary) (d) Determine measurand by using a specific detection system

The determination of an inorganic analyte in an inorganic matrix, e.g. aluminium in rocks, requires the use of classical methods of separation, possibly complexation and a final determination which is designed to remove the effect of interferents by use of a specific chemical reaction(s) or spectrophotometric measurement at a wavelength which is specific to the analyte to be determined. Even so, the ability of this approach to eliminate interference from other elements (or compounds) must be established.

The determination of an inorganic element in an organic matrix usually requires a preliminary treatment to remove the organic matter completely, either by dry-ashing or by oxidation with acids such as nitric, sulfuric or perchloric. Then, the problem reverts to the determination of an inorganic analyte in an inorganic matrix, as above. You should be aware that losses of trace elements can occur during such oxidation processes, either by volatilization or by adsorption onto the surface of the equipment used.

Perhaps the most difficulty arises in the measurement of an organic measurand in an organic matrix because it is then not possible to prevent interference from the matrix by initial destruction of the matrix before carrying out the measurement. This is because the measurand would also be lost during this process. In such a situation, the measurand is first separated from the matrix, usually by solvent extraction.

Table 4.5 Common techniques used to isolate analytes from the sample matrix

Extraction technique	Principle	Analyte/sample matrix
Liquid–liquid extraction	Analyte extracted from liquid sample into an immiscible solvent	Organic compounds in aqueous samples (e.g. pesticides in river water)
Solid-phase extraction (SPE)	Removal of compounds from a flowing liquid sample by retention on a solid sorbent, followed by elution with a solvent	Organic compounds in aqueous samples (e.g. PAHs in waste water)
Soxhlet extraction	Continuous extraction of sample with boiling solvent	Organic compounds in solid samples (e.g. plasticizers in PVC)
Pressurized fluid extraction (PFE)	Sample extracted with solvent at elevated pressure and temperature	Organic compounds in solid samples (e.g. PAHs in soils)
Wet digestion/ashing	Sample heated with strong acid(s)	Metals in organic/inorganic solid samples (e.g. metals in soil samples)
Dry-ashing	Ignition of organic matter in a furnace. Resulting ash dissolved in acid for further analysis	Metals in organic solid samples (e.g. metals in foodstuffs)
Microwave digestion	Sample heated with acids using microwaves under controlled temperature and pressure	Metals in organic/inorganic solid samples (e.g. metals in sediments)

There are a large number of extraction and clean-up techniques. Some of the more common ones are outlined in Table 4.5. The technique selected will depend on the nature of the sample and of the analyte.

Where extraction is incomplete, low results will be obtained. Some methods give results that are only 50% of the true value. In such cases, some workers 'correct' their results by using a recovery factor (see Section 4.6.3). It is always important that such a correction is described in the report that accompanies the result. Where the extraction system is strong enough to remove 90% or more of the measurand from the matrix, it is likely that many other components of the matrix ('co-extractives') will also be present in the extract. This will increase the chances of incorrect results from interferents unless extensive clean-up procedures and a highly selective detection system are subsequently employed.

Regardless of the approach chosen, the procedure should be fully validated to assess whether the recovery of the analyte from the sample matrix is acceptable and to establish that the resulting aliquot submitted for measurement is free from any significant interferences (see Section 4.6).

4.5.5 Losses and/or Degradation

Analytes may be lost at various stages of the analytical procedure for a number of reasons, for example:

- degradation by heat, oxidation;
- losses caused by volatility during digestion, or evaporation;
- losses resulting from adsorption on surfaces, e.g. glassware, crucibles – this is particularly important in trace analysis;
- incomplete extraction of the analyte from the matrix.

In the last case, this may be a physical problem resulting from incomplete penetration by the extraction solvent into the matrix. Alternatively, incomplete recovery of the analyte may result from chemical binding between the analyte and a constituent of the matrix. This is particularly important in the determination of drugs in body tissues where binding to proteins is known to occur. Problems of this kind are documented in the literature. If a new procedure is being developed, it is necessary to investigate the extraction step, e.g. by using radioactive tracers.

4.6 Method Validation

Method validation is defined in the international standard, ISO/IEC 17025 as, *the confirmation by examination and provision of objective evidence that the particular requirements for a specific intended use are fulfilled.* This means that a validated method, if used correctly, will produce results that will be suitable for the person making decisions based on them. This requires a detailed understanding of why the results are required and the quality of the result needed, i.e. its uncertainty. This is what determines the values that have to be achieved for the performance parameters. Method validation is a planned set of experiments to determine these values. The method performance parameters that are typically studied during method validation are selectivity, precision, bias, linearity working range, limit of detection, limit of quantitation, calibration and ruggedness. The validation process is illustrated in Figure 4.2.

If no method exists for the analysis required, then either an existing method has to be adapted or a new method developed. The adapted or developed method will need to be optimized and the controls required identified, hence ensuring that the method can be used routinely in the laboratory. Evidence is then collected so as to demonstrate that the method is 'fit for purpose'. The extent of validation, i.e. the amount of effort that needs to be applied, depends on the details of the problem and the information already available. Figure 4.3 indicates an approach that can be used to decide on the extent of validation required. The answer to DQ 4.2 has already mentioned that the customer may request a particular method. If

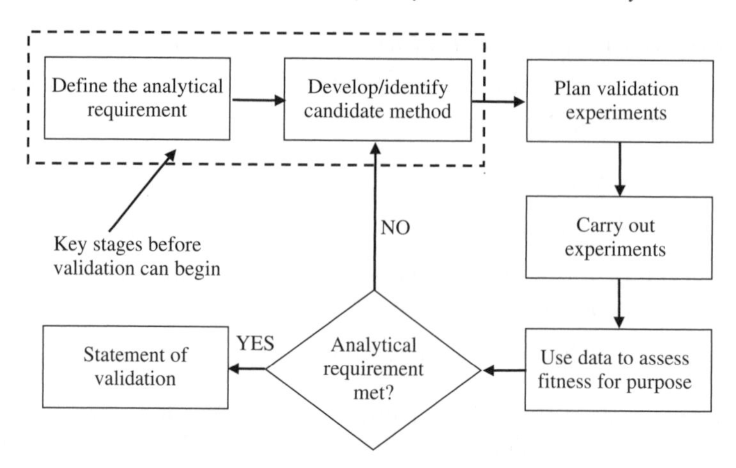

Figure 4.2 Schematic of the method validation process.

this is a published method where the performance characteristics are known, the laboratory has only to confirm their ability to achieve this level of performance. The important parameters, e.g. selectivity, bias, precision and working range, will need to be checked. If the information obtained is satisfactory, the method can be used with confidence. This limited validation where all that is done is confirming that the established published performance can be achieved is called **verification**. The continued satisfactory performance of the method will need to be checked by using adequate control procedures. Where the validation of a standard method is thought to be inadequate (e.g. where an old and poorly validated method is to be used for an important measurement, or where validation data only applies to ideal samples, but the method is to be used for difficult samples), further validation will be required along the lines detailed in the following sections. As a minimum, the laboratory needs to demonstrate that it can meet the requirements specified in the measurement specification.

Ultimately, the amount of validation carried out must ensure that the measurements are fit for their intended use, bearing in mind the level of risk that is acceptable to both the laboratory and its customer. For example, for a highly important measurement there is no option but to rigorously validate the method. For a measurement of medium importance, however, some short-cuts may be taken and the estimate of measurement uncertainty increased to cope with the associated increase in uncertainty. Judgement concerning what is important needs to be made by the laboratory, in collaboration with the customer, where this is not adequately covered in the measurement requirements specification. It must also be recognized that the level of importance can change with time and that additional work may be required where the importance increases. In summary, the extent of validation required will depend on the perceived risk to the laboratory and its customer of inadvertent erroneous data being produced.

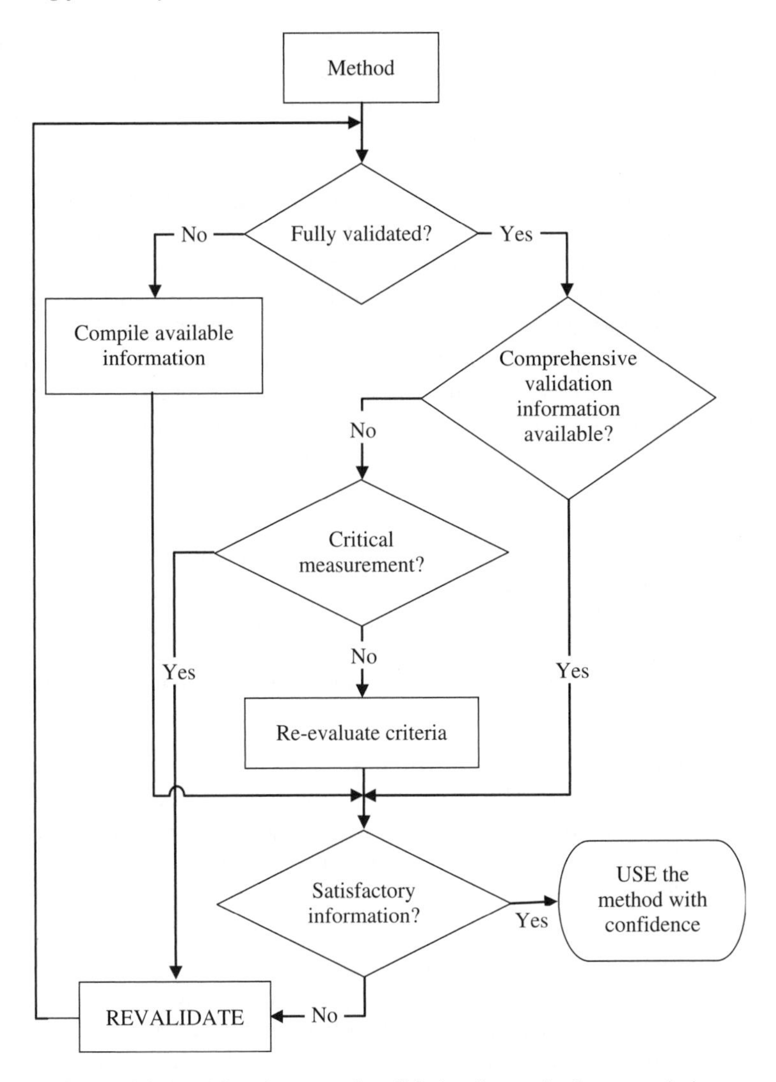

Figure 4.3 Deciding how much validation is required–some choices.

There will be times when a method needs modifying because there is a change in the purchasing policy of the laboratory or to make it suitable for a slightly different application, e.g. change of reagent supplier or use over a wider concentration range. In such situations, full validation is not required; checks need to be made using previously analysed samples or matrix-matched reference materials. The parameters that need to be tested are precision, bias, linearity and, if relevant, the limit of detection. The change may involve replacing one method by another, e.g. change from say GC–FID to GC–MS. In this case, what has to be demonstrated is that the results obtained by using the new method are comparable with those from the established method. Using 'paired comparisons' is the best approach. The samples are split with one portion measured using the old method and the other using the 'new' method. The results are then subjected to statistical analysis.

When no validation data are available, then all of the relevant parameters will have to be studied. The degree of rigour with which the study is carried out will depend on issues such as 'criticality' of the measurement and the availability of validation data on similar methods. There will be cases in the laboratory where a method has been used, satisfactorily, for a long period of time but there is no documentation to demonstrate the performance of the method. It seems unreasonable to require full revalidation when a method has been used successfully for some years. However, the need for objective evidence prevents the validity of such a method being taken for granted. A possible approach is to follow the plan below:

- Identify available information, including information from quality control charts, performance in proficiency testing rounds, literature and validation information on related methods and data concerning comparison with other methods. Use the available information and professional judgement to review each relevant validation issue and sign-off issues adequately addressed and documented.
- Identify significant issues that require further attention and provide the missing information. Unless the validity of the method is in serious doubt, the method can continue to be used. However, the new validation information should be produced over a reasonable time-period and may involve experimental data and professional judgement.

It is important that any method chosen is scientifically sound under the conditions it will be applied. It is also necessary to demonstrate that the equipment, which will be used, is suitable and its use will not influence the results adversely. This includes all types of equipment, e.g. does the volumetric glassware have a suitable tolerance and do the instruments have sufficient sensitivity over the entire range of measurement? The process for demonstrating equipment capability is called 'equipment qualification' and is dealt with in Chapter 5. The staff carrying out validation need to be both qualified and competent in the tasks that they need to carry out.

DQ 4.8

Construct a list of actions you take to demonstrate that a method's performance is adequate for your purpose.

Answer

You will probably have thought of the following:

- repeat measurements;
- bias (recovery) test;
- comparison of results obtained with the candidate method with those from an established validated method or use of alternative detection techniques;
- measurements using materials of known composition, e.g. reference materials;
- participation in proficiency testing schemes or collaborative studies;
- using a number of analysts;
- measurement of samples with low levels of the measurand or none at all (blank value).

The list will probably contain a mixture of processes that lead to values of the performance parameters and quality control checks. A more structured approach will now be taken to method validation. The important performance characteristics are shown in Table 4.6.

There is no relevance to the order of the items in the list. In fact there is no agreed order in which to evaluate the characteristics. Method validation is like putting together a 'jigsaw puzzle'. The more pieces that are in place, the clearer the picture becomes. The difference is that in validation all of the pieces are not always required. Table 4.6 shows when a study of the parameter is required for four different situations. However, it is important that validation is a planned activity or otherwise it can become very labour-intensive and inefficient. Some familiarity with the statistical terms introduced in Chapter 6 is essential before starting out to plan method validation.

Table 4.6 Performance parameters required for validation of different types of analysis

	Type of analysis			
Parameter	Qualitative	Major component	Trace analysis	Physical property
Selectivity/specificity	√	√	√	√
Linearity/working range		√	√	√
Limit of detection	√		√	
Limit of quantitation			√	
Bias/recovery		√	√	√
Precision		√	√	√
Ruggedness	√	√	√	√

You may have noticed that sampling does not appear in Table 4.6. Although sampling is an important issue in chemical analysis, it is not part of method validation. It is assumed that there is sufficient sample available and that the method is validated using materials that have the same or very similar physical and chemical form. Sampling is discussed in detail in Chapter 3.

Associated with method validation, but not part of it, are two properties of results that have been previously mentioned. These parameters are *measurement uncertainty* and *metrological traceability*. Measurement uncertainty is covered in Chapter 6 and metrological traceability in Chapter 5. If considered at the planning stage of method validation, the information obtained during validation is a valuable input into measurement uncertainty evaluation. Traceability depends on the method's operating procedures and the materials being used.

4.6.1 Selectivity

During method development, it will have been established that the method is capable of measuring the measurand of interest. However, part of the aim of method validation is to verify that *only* the measurand of interest is actually measured. The extent to which a method can unambiguously detect and determine a particular analyte in a mixture, without interference from the other components in the mixture, is referred to as *selectivity* or *specificity*. In some fields of measurement, the terms are used interchangeably and this may cause confusion. Selectivity is the term recommended for use in analytical chemistry to express the extent to which a particular method can be used to determine analytes, under given conditions, in the presence of components of similar behaviour [6]. Selectivity will be enhanced by measuring a unique property, such as absorbance at a specific wavelength, and by separating the analyte from other substances present in the sample.

If it has not been adequately addressed during method development, study the selectivity by analysing samples ranging from pure measurement standards spiked with potential interferents, to known mixtures that match 'real-sample' compositions. Serious interferences need to be eliminated, but minor effects can be tolerated and included in the estimation of method bias and its associated uncertainty.

For complex sample types, if there is any doubt concerning the ability of the method to unambiguously identify and measure the analyte of interest, check the method using a closely matched matrix reference material, or check the sample using an alternative validated method.

4.6.2 Precision

Precision and bias both influence a result – this is illustrated in Figure 4.4 and discussed below. As mentioned in Section 4.3.2, the precision of a method is a statement of the closeness of agreement between independent test results obtained

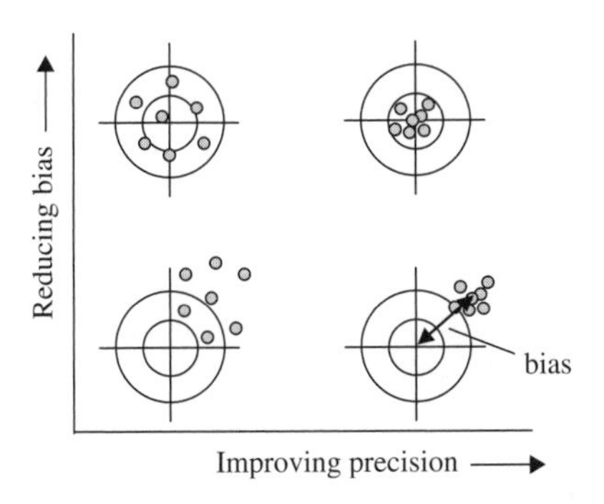

Figure 4.4 Precision and bias.

under stated conditions. The precision is usually stated in terms of the standard deviation (s), the relative standard deviation (RSD), sometimes called the coefficient of variation (CV), or the standard deviation of the mean (SDM) of a number of replicates. The equations for calculating these parameters are given in Chapter 6. The value of the precision depends on a variety of factors, including the number of parameters that are varied during the precision study and the level of variation in the operating conditions, as illustrated in Figure 4.5.

Clearly, precision studies should mirror the operating conditions used during the routine use of the method. Individual sources of imprecision, such as change of operator or instrument type, can be studied but two types of overall precision are commonly estimated, i.e. repeatability and reproducibility.

Repeatability is a measure of the short-term variation in measurement results and is the precision that can be most easily determined. It is often used to establish compliance with method performance criteria. While repeatability is a

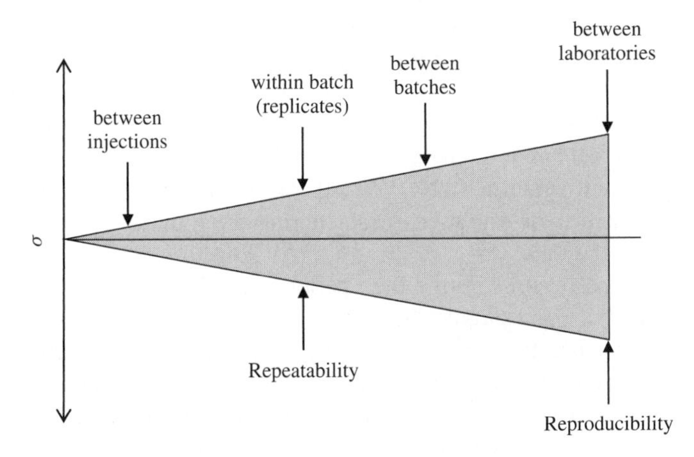

Figure 4.5 Effect of varying conditions during precision studies.

useful performance parameter, on its own it cannot indicate the spread of results that can be expected over the longer term. **Reproducibility** is a type of precision relating to a situation where environmental conditions and other factors will have changed and usually the results are obtained in different laboratories and at different times. **Intermediate precision** or within-laboratory reproducibility is the term often used to describe data obtained over an extended period in a single laboratory. Inter-laboratory reproducibility studies, carried out over a period of months, invariably yield larger precision values, as a result of greater variation in the measurement conditions. It has already been mentioned that the value of the reproducibility of a method is often greater than the repeatability, typically by a factor of two to three.

The greater the number of repeat measurements used to estimate the precision, the greater is the confidence that can be placed in the estimate, but there is little gained from making more than about fifteen repeat measurements. The measurements need to be 'independent' of each other. For example, independent portions of the sample need to be weighed out, dissolved, extracted, etc. It is not normally sufficient to make repeat measurements of the same prepared test solution. The question is – what is the minimum number of repeats required? To estimate both the method precision and the measurement precision, replicate measurements on several independent test portions are required. For most purposes, duplicate (or preferably, triplicate) measurements of between seven and fifteen independent test portions will suffice. Where much of the imprecision is in the final measurement stage, then the work load can be reduced, without seriously impairing the quality of the precision estimate, by reducing the number of test portions and increasing the number of replicate measurements made on each test solution. Sometimes, data will be available from different precision experiments carried out on different samples and on different occasions. As long as the variances are not statistically significantly different, it is possible to 'pool' the data and calculate a 'pooled' standard deviation [9].

Within-laboratory reproducibility studies should cover a period of three or more months and these data may need to be collected during the routine use of the method. It is possible, however, to estimate the intermediate precision more rapidly by deliberately changing the analyst, instrument, etc. and carrying out an analysis of variance (ANOVA) [9]. Different operators using different instruments, where these variations occur during the routine use of the method, should generate the data.

Precision studies should mirror the operating conditions used during routine use of the method. For example, the range of operating conditions, such as the variation in the laboratory temperature, needs to reflect that which will occur in practice. In addition the same number of replicate measurements per test portion should be used. Where a range of analytes is measured by a single method (e.g. pesticides by GC or trace elements by ICP–MS), or where different matrix types are encountered, it is necessary to determine the precision parameters for representative situations. The dependence of the precision on the analyte concentration should be determined. The standard deviation may not be constant over a range of concentrations but often the *RSD* (*CV*) is roughly constant over a wider range. In Table 4.3, the acceptable levels for the coefficient of variation were different for the different ranges of concentration and this was linked to the Horwitz equation. A plot of this function is shown in Figure 4.6 [10].

The equation representing this curve was introduced in Section 4.4 (equation (4.4)). However, a more contemporary model based on results from Proficiency Testing schemes has shown that the relationship is best represented if three equations are used to cover from high to low concentrations, as shown in

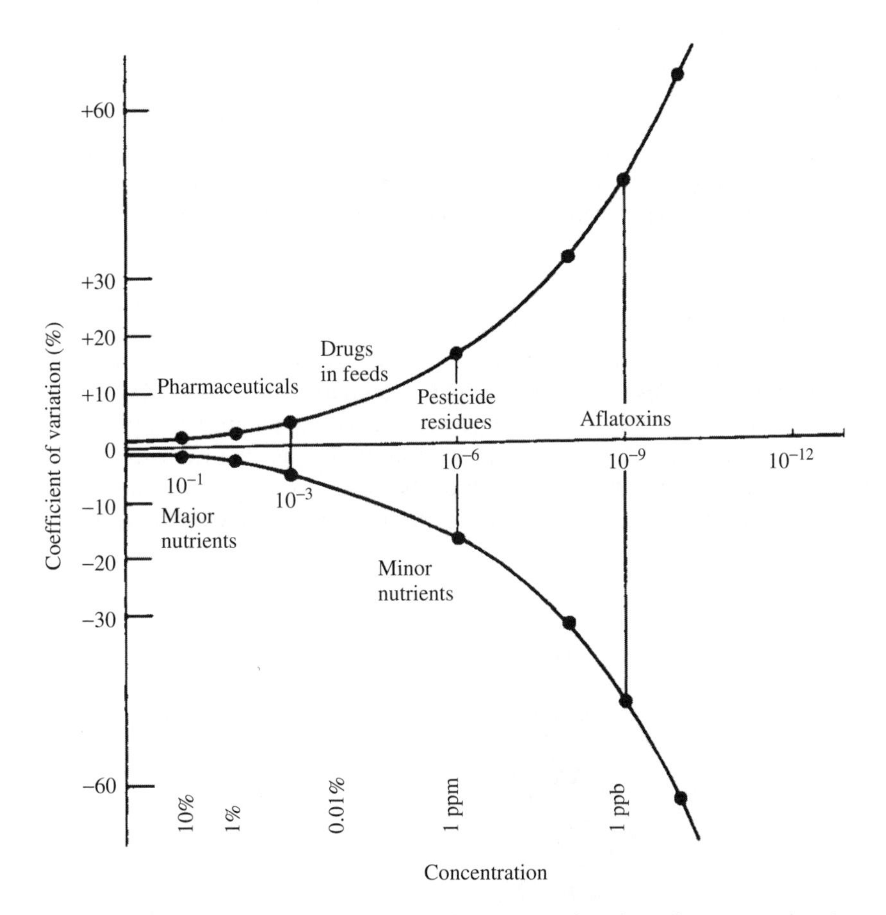

Figure 4.6 Interlaboratory coefficient of variation as a function of concentration (note that the filled circles are values calculated by using equation (4.4), *not* experimental points) [10]. Reproduced by permission of AOAC International, from Horwitz, W., *J. Assoc. Off. Anal. Chem.*, **66**, 1295–1301 (1983).

equation (4.8) [11]. The middle range is the original equation:

$$s_R = \begin{cases} 0.22c, & \text{if } c < 1.2 \times 10^{-7} \\ 0.02c^{0.8495}, & \text{if } 1.2 \times 10^{-7} \leq c \leq 0.138 \\ 0.01c^{0.5}, & \text{if } c > 0.138 \end{cases} \quad (4.8)$$

Precision estimates are key method performance parameters and are also required in order to carry out other aspects of method validation, such as bias and ruggedness studies. Precision is also a component of measurement uncertainty, as detailed in Chapter 6. The statistics that are applied refer to random variation and therefore it is important that the measurements are made to comply with this requirement, e.g. if change of precision with concentration is being investigated, the samples should be measured in a random order.

SAQ 4.2

Using the appropriate equation, calculate the expected %*CV* value for the analysis of samples of milk powder containing aflotoxins at a concentration of 0.562 μg kg^{-1}. Compare your value with that obtained from Figure 4.6.

4.6.3 Bias/Trueness

As mentioned in Section 4.3.3, bias is the difference between the mean value ($\overline{x}$) of a number of test results and an accepted reference value (x_0) for the test material. As with all aspects of measurement, there will be an uncertainty associated with any estimate of bias, which will depend on the uncertainty associated with the test results $u_{\overline{x}}$ and the uncertainty of the reference value u_{RM}, as illustrated in Figure 4.7. Increasing the number of measurements can reduce random effects

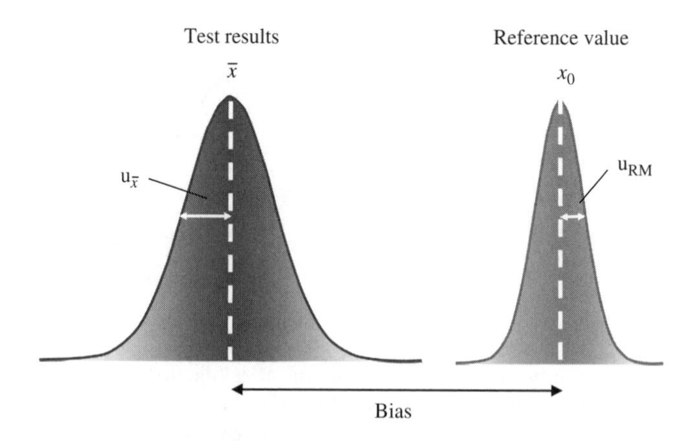

Figure 4.7 Illustration of bias.

(improve precision) but systematic effects (bias) cannot be reduced. The latter have to be removed or taken into account.

The overall bias is usually made up of a number of components. The chemical interferences discussed in Section 4.5.4 are just one potential source of measurement bias. Other causes of bias include the following: matrix effects (such as changes in acid strength or viscosity) that can enhance or suppress the measurement signal; measurement equipment bias, such as blank signals, or non-linearity; and incomplete recovery of the analyte from the sample matrix. Thus, there can be a number of bias effects, which can be positive or negative. Some of the effects are associated with the sample being analysed while some are associated with the method and some with the particular laboratory doing the work. For a measurand that is defined by the method of measurement (empirical methods, such as the analysis of fibre), the method bias is by definition zero, but there may also be a laboratory bias. Individual effects can be studied separately, or more commonly the overall bias effect can be evaluated as a measurement bias.

Specific bias effects, which could be large and need to be understood and minimized, should have been studied during method development, e.g. the presence or absence of a potential interfering substance, or the effect of changing a derivatization reaction time. The task during validation is to study residual effects and this can be done as part of ruggedness studies, as discussed in Section 4.6.5.

Measurement bias is determined by comparing the mean of measurement results obtained for a reference material, using the method being validated, with the assigned property values for that reference material. The number of replicate analyses required (n) depends on the precision of the method (s) and the level of bias (δ) that needs to be detected [12]. A useful approximation is shown in the following equation:

$$n = 13 \times (s/\delta)^2 + 2 \quad (4.9)$$

As a 'rule of thumb', at least seven replicates should be carried out. With this number of replicates a bias of about twice the standard deviation of the method will be detected. If the method has a large scope in terms of concentration and/or matrix type, one should use a number of independent reference materials covering the expected measurement range and sample type. An ideal reference material is one which is certified (a Certified Reference Material (CRM)) and closely matches the sample with regards to sample form, matrix composition and concentration of analyte [13]. Where well-matched matrix CRMs are available, they should be used for evaluating bias. Unfortunately, ideal materials are seldom available and a combination of the best available matrix reference materials and 'spiking experiments' is the best option. Spiking experiments involve the analysis of real samples before and after adding, 'spiking', known amounts of pure analyte. For an unbiased method, the difference between the two results would equal the amount of added analyte, within the uncertainties of measurement, as illustrated in Figure 4.8. However, gravimetric spiking only gives reliable bias estimates when

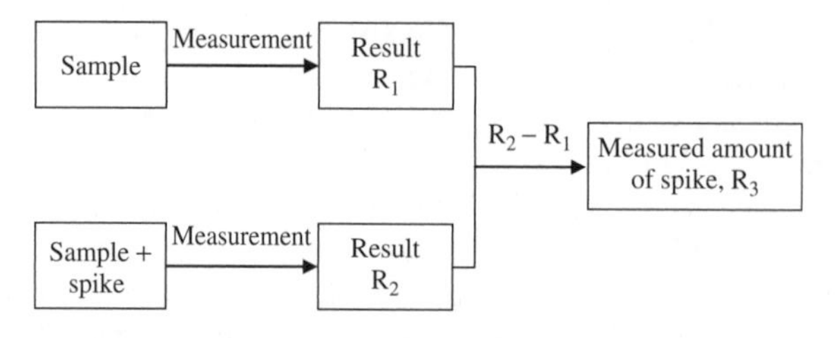

Figure 4.8 Evaluating bias by using gravimetric spiking.

the naturally occurring and spiked analytes are in equilibrium. For example, native analyte may be bound tightly within the matrix, whereas the spiked analyte may be loosely adsorbed on the surfaces of the sample particles. In addition whereas equilibrium may be achieved at high concentrations, this may not be the case at trace levels. While this is not a problem if the samples are totally dissolved, bias can occur if extraction from a solid is a step in the method. Care should be taken to ensure that the matrix composition is not changed significantly during spiking and that the concentration of spiked materials closely matches those of the samples to be analysed. Establishment of equilibrium conditions is important and this can be achieved by leaving the spiking solution in contact with the sample for several hours, possibly overnight, and controlling properties such as particle size. Whenever possible the solvent used to add the spike should be different from that used in any subsequent extraction. The effect of analyte concentration on the bias needs to be assessed, as does any concentration effect on the behaviour of native and spiked material.

Measurement bias can also be determined by comparing the results obtained using the method of interest with the results obtained using a reference method of known bias. This approach is very similar to comparison with a reference material, with the sample subject to measurement acting as a transient reference material.

Based on the study results, decide how frequently and how rigorously the bias needs to be checked during routine use of the method (e.g. every batch using a single spike) and document it in the method as part of quality control.

4.6.3.1 Calculating Bias Parameters

The measurement bias, B, can be calculated as the ratio (often expressed as a percentage) of the difference between the mean of a number of determinations of a test sample, obtained under repeatability conditions, and the 'true' or accepted concentration for that test sample, as shown in the following equation:

$$\%B = \frac{(\overline{x} - x_0)}{x_0} \times 100 \qquad (4.10)$$

where $\overline{x}$ is the mean value obtained from repeat measurement of a reference material using the method in question and x_0 is the assigned property value for the reference material.

Measurement bias is often called 'recovery', which can be expressed as a ratio (R), or a percentage (%R) and can be greater or less than 100%. In some fields of measurement, recovery refers to the amount of added (spiked) analyte recovered during analysis (equation (4.11)). In other fields, recovery is taken as an estimate of the proportion of the total analyte (native plus any added spike) present in a sample that is measured (recovered) by the method. The relationship between recovery and bias is shown in equation (4.12):

$$\%\text{R} = \frac{\overline{x}}{x_0} \times 100 \quad (4.11)$$

$$\%\text{R} = 100 + \%\text{B} \quad (4.12)$$

Where recovery has been studied by spiking, the %recovery is calculated from the following equation:

$$\%\text{R} = \frac{(\text{C}_{\text{sp}} - \text{C}_{\text{b}})}{\text{C}_{\text{s}}} \times 100 \quad (4.13)$$

where C_{sp} is the mean value obtained from repeat measurement of a sample, after the addition of a spike, C_b is the mean value obtained from repeat measurement of a sample, prior to spiking, and C_s is the calculated increase in concentration of the sample after adding the spike.

Where there is a statistically significant bias, routine measurement results can be corrected as shown in the following equation:

$$\text{C}_{\text{cor}} = \frac{\text{C}_{\text{obs}}}{\%\text{R}} \times 100 \quad (4.14)$$

where C_{cor} is the corrected measurement result for a test sample and C_{obs} is the observed measurement result for a test sample.

SAQ 4.3

The certified value for cholesterol in a Certified Reference Material is 274.7 ± 9.0 mg $(100\,\text{g})^{-1}$. The results of repeat experiments to measure the recovery of cholesterol from the CRM are as follows: 271.4, 266.3, 267.8, 269.6, 268.7, 272.5, 269.5, 270.1, 269.7, 268.6 and 268.4. The results are reported in mg $(100\,\text{g})^{-1}$. Calculate the bias of the method, the percentage bias and the percentage recovery.

4.6.4 Measurement Range, Limit of Detection (LoD) and Limit of Quantitation (LoQ)

During validation, the relationship between response and concentration is established. Checks also are made to ensure that the linearity of the method does not make too large a contribution to the measurement uncertainty (the uncertainty due to the calibration should contribute less than about 20% of the largest uncertainty component). The calibration schedule required during routine operation of the method is also established. It is wise to carry out sufficient checks on some or all of the following performance parameters, in order to establish that their values meet any specified limits.

Sensitivity is the rate of change of the measuring instrument response with change in concentration. This is better known as the *slope* of the calibration graph. Clearly, the greater the sensitivity, the better the method is able to distinguish between similar concentrations, as a small difference in concentration will lead to a large difference in observed response. The sensitivity may change with concentration, as illustrated in Figure 4.9, but the calibration graph can often be expected to be linear over a wide range of concentrations.

The **linearity and working range** of a method are determined by examining samples with different analyte concentrations and determining the concentration range for which acceptable calibration linearity and measurement uncertainty are achieved. Linearity can be assessed by visual inspection of a plot of measured responses against standard concentrations, or by statistical measures. The calibration response does not have to be perfectly linear for a method to be useable. All that is required is the equation relating response to concentration – this is known as the *calibration function*. The working range is the region where the results will have acceptable uncertainty and can be greater than the linear range. The lower end of the working range is defined by the limit of quantitation (LoQ)

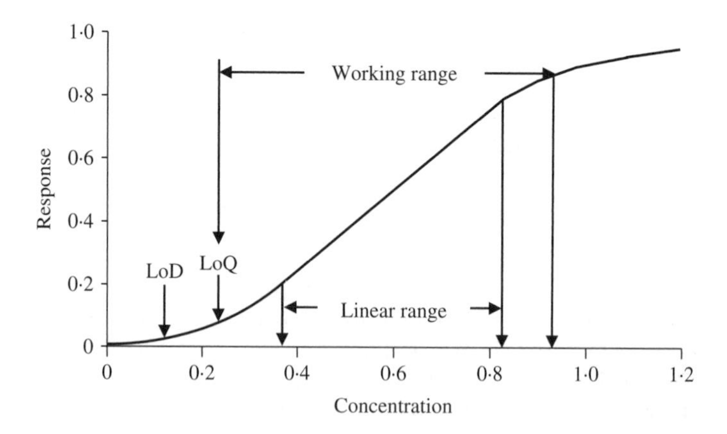

Figure 4.9 Illustration of some key performance parameters.

and the upper end by the point where there is insufficient change of response per unit change of concentration, as shown in Figure 4.9. The linear range can vary with matrix type and thus it may need to be checked for a number of sample types.

The **limit of detection** (LoD) has already been mentioned in Section 4.3.1. This is the minimum concentration of analyte that can be detected with statistical confidence, based on the concept of an adequately low risk of failure to detect a determinand. Only one value is indicated in Figure 4.9 but there are many ways of estimating the value of the LoD and the choice depends on how well the level needs to be defined. It is determined by repeat analysis of a blank test portion or a test portion containing a very small amount of analyte. A measured signal of three times the standard deviation of the blank signal ($3s_{bl}$) is unlikely to happen by chance and is commonly taken as an approximate estimation of the LoD. This approach is usually adequate if all of the analytical results are well above this value. The value of s_{bl} used should be the standard deviation of the results obtained from a large number of batches of blank or low-level spike solutions. In addition, the approximation only applies to results that are normally distributed and are quoted with a level of confidence of 95%.

The UK water industry uses $\text{LoD} = 2\sqrt{2}ts_w$, where s_w is the 'within-batch' standard deviation of the blank results, and t is the Student t-statistic for the number of degrees of freedom assigned to the standard deviation [2]. This is approximately $4.65s_w$ and requires that a single blank correction will be applied to responses for test samples from a single batch. The LoD obtained during validation studies is an indication of the 'fitness for purpose' of the method. It should be checked regularly during the use of the method to ensure that there are no baseline changes.

DQ 4.9

How do you distinguish between the instrument detection limit (IDL) and the method detection limit (MDL)?

Answer

The IDL is an instrument parameter and is the lowest concentration of the measurand that results in an instrument response, reliably. This can be obtained from measurements of pure analyte. This is in contrast to the MDL (LoD) which is based on measurements of a blank real sample or a low-level spike that has been processed through all of the steps of the method. Clearly, it is the latter that is relevant for test samples. The IDL can also be estimated from the instrument signal-to-noise ratio; it is approximately three times this ratio. In this case, the value obtained has to be converted to concentration units.

Table 4.7 Responses from ten replicate analyses at different concentration levels

Concentration ($\mu g\ g^{-1}$)	Positive/negative
200	10/0
100	10/0
75	5/5
50	1/9
25	0/10
0	0/10

For tests designed to detect the presence or absence of an analyte, the threshold concentration that can be detected can be determined from replicate measurements over a range of concentrations. These data can be used to establish at what concentration a cut-off point can be drawn between reliable detection and non-detection. At each concentration level, it may be necessary to measure approximately ten replicates. The cut-off point depends on the number of false negative results that can be tolerated. It can be seen from Table 4.7 that for the given example the positive identification of the analyte is not reliable below 100 $\mu g\ g^{-1}$.

The **limit of quantitation** (LoQ) is the lowest concentration of analyte that can be determined with an acceptable level of uncertainty. This should be established by using an appropriate reference material or sample. It should not be determined by extrapolation. Various conventions take the approximate limit to be 5, 6 or 10 times the standard deviation of a number of measurements made on a blank or a low-level spiked solution.

The approaches described above give approximate values for the LoD and LoQ. This is sufficient if the analyte levels in test samples are well above the LoD and LoQ. If the detection limits are critical, they should be evaluated by using a more rigorous approach [1, 2, 14]. In addition, the LoD and LoQ sometimes vary with the type of sample and minor variations in measurement conditions. When these parameters are of importance, it is necessary to assess the expected level of change during method validation and build a protocol for checking the parameters, at appropriate intervals, when the method is in routine use.

4.6.4.1 Linearity Checks

Linearity is a measure of a method's ability to give a response that is directly proportional to the concentration of the analyte being studied. Sometimes the response needs some transformation (via a mathematical function) before linearity can be assessed. A quantitative evaluation of linearity may be obtained from various statistics related to fitting linear or higher functions to the data.

Establishing linearity will normally require more standards (and more replication at each concentration) than is typical for calibration of a validated method in regular use. However, note that when establishing linearity the solutions of known concentration do not need to be independent. Linearity needs to cover the method scope in terms of analyte and matrix combinations and should cover a concentration range that is greater than the expected range for samples by, for example, ± 20%.

For a reliable linearity study:

- Distribute the concentration values (x) of the calibrating solutions evenly over the range of interest. One or two values at the extreme ends can cause 'leverage' of the line [15].
- Study a minimum of six concentration levels. Measure these in random order.
- Measure two to five replicates at each level.
- Ensure that the responses from samples are close to the mean response, $\overline{y}$, of the calibration set. This will decrease the error contribution from the least-squares estimate of the regression line.
- Pure materials can be used to assess the linearity of an instrument response.
- Matrix-matched Certified Reference Materials or spiked samples should be used to determine the linearity of a method.
- Assess the linearity by visual inspection of the graph of response against concentration.
- Check for the presence of outliers. If there are suspect values, check by using a statistical test, either the Grubbs or Dixon tests [9]. Do not reject possible outliers just on the basis of statistics.
- Other statistical parameters that can be used include examination of residuals and the output from the ANOVA table of regression statistics. This may indicate that a non-linear response function should be checked [9].

A detailed treatment of linearity evaluation is beyond the scope of this present book but a few general points are made below. It is important to establish the homogeneity of the variance ('homoscedasticity') of the method across the working range. This can be done by carrying out ten replicate measurements at the extreme ends of the range. The variance of each set is calculated and a statistical test (F test) carried out to check if these two variances are statistically significantly different [9].

Linearity should always be assessed initially by visual inspection of the plotted data and then by statistical evaluation. The linearity of the instrument response needs to be established because without this information it is difficult to attribute causes of non-linearity. The supporting statistical measures include correlation coefficient (r, r^2, etc), residual plot, residual standard deviation and significance tests for the slope and intercept. It should be remembered that the correlation coefficient, r, is not a measure of linearity – it just indicates the extent of the relationship between the dependent and independent variables, e.g. the instrument response and analyte concentration. It is important to carry out a visual inspection of the residual plot to check for trends. The linearity of a method needs to be characterized in order to establish the calibration protocol for future use of the method.

4.6.5 Ruggedness Testing

Ruggedness testing evaluates how small changes in the method conditions affect the measurement result, e.g. small changes in temperature, pH, flow rate, composition of mobile phase, etc. The aim is to identify and, if necessary, better control method conditions that might otherwise lead to variation in measurement results, when measurements are carried out at different times or in different laboratories. It can also be used to improve precision and bias.

Ruggedness testing can be carried out by considering each effect separately, by repeating measurements after varying a particular parameter by a small amount (say 10%) and controlling the other conditions appropriately. However, this can be labour-intensive as a large number of effects may need to be considered. Since for a well-developed method, most of the effects can be expected to be small, it is possible to vary several parameters at the same time. A number of experimental design approaches are available in the literature [16, 17].

Any stable and homogeneous sample within the scope of the method can be used for ruggedness-testing experiments. Youden and Steiner [16] describe a simple design (the Plackett–Burman design [17]) which allows seven independent factors to be examined in eight experiments. Factors that might be thought to influence a result could include, acid concentration, time of extraction, flow rate, temperature, etc. If one can identify two levels of each of the parameters under investigation, there are 128 combinations that can be written out. However, the design described allows information to be gathered from only eight experiments. Let A, B, C, D, E, F and G denote the nominal levels of the seven chosen parameters and a, b, c, d, e, f and g represent the alternative levels. The chosen levels may be the extreme values of the parameter, e.g. the two extremes of temperature likely to be encountered during use of the method or the two extremes of water hardness likely be present, etc. In Table 4.8, you will see that the selected eight combinations of these letters leads to a balance between the upper- and lower-case letters. This table shows the values, for the seven factors, to be used when running the eight experiments. The results from the experiments are shown as l, m, p, w, v, x, y and z. For each factor, the difference is calculated between the average of the results obtained with the factor at its nominal value and the average of the results obtained with the factor at its alternative value.

Table 4.8 The Plackett–Burman design [17]

Experiment number	Method parameter							Observed result
1	A	B	C	D	E	F	G	l
2	A	B	c	D	e	f	g	m
3	A	b	C	d	E	f	g	p
4	A	b	c	d	e	F	G	w
5	a	B	C	d	e	F	g	v
6	a	B	c	d	E	f	G	x
7	a	b	C	D	e	f	G	y
8	a	b	c	D	E	F	g	z

To find if changing factor 'A' to 'a' has an effect, Δ_A is calculated as shown in the following equation:

$$\Delta_A = \frac{l + m + p + w}{4} - \frac{v + x + y + z}{4} \tag{4.15}$$

Experiments 1–4 had the factor at the nominal level, 'A', while experiments 5–8 had the factor at the alternative level, 'a'. Inspection of Table 4.8 shows that with this combination the effect of the other factors cancels out. The seven factors can be dealt with in a similar way by grouping the nominal level for that factor and subtracting the alternative level. Changing factor 'B' to 'b' is examined by calculating Δ_B, as follows:

$$\Delta_B = \frac{l + m + v + x}{4} - \frac{p + w + y + z}{4} \tag{4.16}$$

The next step is to arrange the seven differences, Δ_A to Δ_G, in numerical order (ignoring the sign). To calculate if any of the differences are statistically significant, a statistical test (t-test) is applied. Equation (4.17) is used to compare the difference $|\Delta_i|$ with the expected precision of the method, s. The value of t used corresponds to the value obtained from statistical tables for the degrees of freedom appropriate for the estimation of s and the level of confidence used. For example, if the method standard deviation was obtained from ten results, i.e. nine degrees of freedom, $t(95\%) = 2.262$.

$$|\Delta_i| > \frac{ts}{\sqrt{2}} \tag{4.17}$$

For cases where equation (4.17) is true, the change from the nominal to the alternative level is significant. However, the results of the test will be misleading if the factors investigated are not independent. Such a study may be used to set the level of control that should be applied at particular stages of the method, e.g. adjust the pH to 6.5 ± 0.2. It is also possible to study the effect of potential interferences by using this approach.

4.6.6 'Sign-off' and Documentation

Method validation is carried out to provide objective evidence that a method is suitable for a given application. A formal assessment of the validation information against the measurement requirements specification and other important method performance parameters is therefore required. Although validation is described as a sequential process, in reality it can involve more than one iteration to optimize some performance parameters, e.g. if a performance parameter is outside the required limits, method improvement followed by revalidation is needed.

Method validation provides information concerning the method's performance capabilities and limitations, when applied under routine circumstances and when it is within statistical control, and can be used to set the QC limits. The warning and action limits are commonly set at twice and three times the within-laboratory reproducibility, respectively. When the method is used on a regular basis, periodic measurement of QC samples and the plotting of these data on QC charts is required to ensure that the method is still within statistical control. The frequency of QC checks should not normally be set at less than 5% of the sample throughput. When the method is new, it may be set much higher. Quality control charts are discussed in Chapter 6.

The validation is therefore not complete until there is a detailed description of the method and records of the validation study. A responsible person needs to 'sign' that the method meets the requirements, i.e. it is 'fit for purpose'. The documentation facilitates the consistent application of the method, within its scope and defined performance parameters. This, in turn, helps ensure that when the method is applied in different laboratories or at different times, the measurand is the same and that the measurement results are comparable. Documentation is also required for quality assurance, regulatory and contractual purposes.

Changes to methods will occur due to changes in applications and technological developments and these changes need to be formally implemented and recorded, after appropriate validation. A periodic review (commonly every two years) and, where necessary, revision is considered to be 'good practice'.

A system of document control, including restricting unofficial copying, is required, in order to ensure that only authorized methods are used and that these are the latest revisions. This requires attention to details, such as revision number, who authorized the method, who has amended it, issue date, date of next revision, number of copies and who owns them, etc. A variety of documentation formats are available, but a standard format for method documentation is provided in ISO 78-2:1999 and a layout based on this Standard is provided in the Appendix to this chapter [18].

Summary

The suitability of a method to solve a customer's problem will depend both on the method used and the way it is carried out. This chapter takes the analyst through the process of selecting a method for a particular purpose and provides some of the major sources of published methods. It describes the parameters of the method which define its performance, i.e. precision, bias, working range and ruggedness. These are the parameters studied during method validation. The process of method validation is described. Reasons are given why sometimes incorrect results are obtained. Some of these causes are easier to solve than others, while sometimes it is just because the method has not been written down in an unambiguous form. This chapter also includes the layout of method documentation based on an International Standard.

References

1. Currie, L., *Pure Appl. Chem.*, **67**, 1699–1723 (1995).
2. Cheeseman, R. V. and Wilson, A. L. (revised by Gardner, M. J.), *NS30, A Manual on Analytical Quality Control for the Water Industry*, ISBN 0902156-853, Water Research Centre plc, Marlow, UK, 1989.
3. 'Accuracy (Trueness and Precision) of Measurement Methods and Results – Part 1. General Principles and Definitions', ISO 5725-1:1994, International Organization for Standardization (ISO), Geneva, Switzerland 1994.
4. 'Statistics – Vocabulary and Symbols – Part 2. Applied Statistics', ISO 3534-2:2006, International Organization for Standardization (ISO), Geneva, Switzerland, 2006.
5. 'International Vocabulary of Metrology – Basic and General Concepts and Associated Terms (VIM)', ISO/IEC Guide 99:2007, International Organization for Standardization (ISO), Geneva, Switzerland, 2007.
6. Vessma, J., Stefan, R. I., van Staden, J. F., Danzer, K., Lindner, W., Burns, D. T., Fajgelj, A. and Muller, H., *Pure Appl. Chem.*, **73**, 1381–1386 (2001).
7. Albert, R. and Horwitz, W., *Anal. Chem.*, **69**, 789–790 (1997).
8. Ardrey, R. E., *Liquid Chromatography–Mass Spectrometry: An Introduction*, AnTS Series, ISBN 0-471-49801-7, AnTS Series, John Wiley & Sons, Ltd, Chichester, UK, 2003.
9. Miller, J. N. and Miller, J. C., *Statistics and Chemometrics for Analytical Chemistry*, 5th Edition, ISBN 0-131-291920, Pearson Education Ltd, Harlow, UK, 2005.
10. Horwitz, W., *J. Assoc. Off. Anal. Chem.*, **66**, 1295–1301 (1983).
11. Thompson, M., *Analyst*, **125**, 385–386 (2000).
12. Becker, D., Christensen, R., Currie, L., Diamondstone, B., Eberhardt, K. R., Gills, T., Hertz, H., Klouda, G., Moody, J., Parris, R., Schaffer, R., Steel, E., Taylor, J., Watters, R. and Zeisler, R., *Use of NIST Standard Reference Materials for Decisions on Performance of Analytical Chemical Methods and Laboratories*, NIST Special Publication 829, National Institute of Standards and Technology (NIST), Gaithersburg, MD, USA, 1992.
13. Barwick, V., Burke, S., Lawn, R., Roper, P. and Walker, R., *Applications of Reference Materials in Analytical Chemistry*, ISBN 0-85404-448-5, The Royal Society of Chemistry, Cambridge, UK, 2001.
14. Kateman, G. and Buydens, L., *Quality in Analytical Chemistry*, 2nd Edition, ISBN 0-471-55777-3, John Wiley & Sons, Ltd, Chichester, UK, 1993.
15. Barwick, V., *Preparation of Calibration Curves. A Guide to Best Practice*, Valid Analytical Measurement (VAM) Programme, LGC, Teddington, UK, 2003. [http://www.nmschembio.org.uk] (accessed 11 November, 2007).
16. Youden, W. J. and Steiner, E. H., *Statistical Manual of the Association of Official Analytical Chemists*, ISBN 0-935584-15-3, Association of Official Analytical Chemists (AOAC), Arlington, VA, USA, 1975.
17. Plackett, R. L. and Burman, J. P., *Biometrika*, **33**, 305–325 (1946).
18. 'Chemistry–Layouts for Standards – Part 2: Methods of Chemical Analysis', ISO 78-2:1999, International Organization for Standardization (ISO), Geneva, Switzerland, 1999.

Appendix: Layout for Method Documentation

The following layout is based on ISO 78-2 [18]. Note that method documentation requires appropriate control (see Section 4.6.6). In addition to the headings described below, each page of the method should be marked with the page number, the total number of pages and the version number.

0 Update and Review Summary

This section has a twofold purpose. First, it is intended to enable minor changes to be made to the text of the method without the need for a full revision and reprint of the method. Secondly, it is recommended that every method should be reviewed periodically for 'fitness for purpose' and the summary serves as a record that this has been done. The summary typically would be located at the front of the method, just inside the front cover.

0.1 Updates

Hand-written changes to the text of the method are acceptable provided the changes are also recorded in the table below (hand-written entries acceptable) and appropriately authorized. It is implicit that the authorization endorses the fact that the effects of the changes on the method validation have been investigated and cause no problem, and that the changes have been made to all copies of the method.

#	Section	Nature of Amendment	Date	Authorization
1 (e.g.)	3.4	Change flow rate to 1.2 ml min^{-1}	8/2/06	FEP

0.2 Review

Two years has been suggested as a suitable interval.

Review Date	Outcome of Review	Next Review Date	Authorization

1 Title

Preferred format: Determination of A{*analyte or measurand*} (in the presence of B{*interference*}) in C{*matrix*} using D{*principle*}.

2 Warning and Safety Precautions

Detailed precautions may be given in the relevant sections, but notice must be drawn here to the existence of hazards and the need for precautions. Include nil returns.

Provide suitable warnings of any hazards involved with:

- handling the samples;
- handling or preparing solvents, reagents, standards or other materials;
- operation of equipment;
- requirements for special handling environments, e.g. fume cupboards;
- consequences of scaling-up experiment (explosion limits);
- disposal of any items/materials used.

3 Scope

This section enables a potential user to see quickly whether the method is likely to be appropriate for the desired application. The following details should be covered:

- the analyte(s) which can be determined by the method;
- the form in which analyte(s) are determined – speciation, total/available, etc.;
- the sample matrices within which those analyte(s) may be determined;
- the concentration range of analyte(s) over which the method may be used;
- known interferences which prevent or limit the working of the method;
- the technique used by the method;
- the minimum sample size;
- client related issues, such as suitability for establishing compliance with specification or regulation requirements.

4 Normative References

List the references where compliance with their requirements is essential.

5 Definitions

Define any unusual terms; use ISO definitions wherever possible. Quote sources. Chemical structures can be included here if relevant.

6 Principle

Outline the principle by which the analytical technique operates. A flow-chart may help. This section should be written so as to allow an 'at-a-glance' summary of how the method works. Include an explanation of the principle of the calculations. Where appropriate to clarifying the working of the method or calculations, include details of any relevant chemical reactions (for example, this may be relevant where derivatization is involved, or titrimetry).

For example, 'The concentration is derived from a 6 point calibration curve by reading off the concentration, corresponding to the sample absorbance, corrected for the blank value, and multiplying it by the concentration factor'.

7 Reagents and Materials

List all of the reagents, materials, blanks, QC samples, standards and Certified Reference Materials required for the analytical process, numbered for later reference. List the following:

- details of any associated hazards, including instructions for disposal;
- analytical grade;
- need for calibration and QC materials to come from independent batches;
- details of preparation, including need to prepare in advance;
- containment and storage requirements;
- shelf-life of raw materials and prepared reagents;
- required concentration, noting the units;
- labelling requirements.

8 Apparatus and Equipment

Describe individual items of equipment and how they are connected in sufficient detail to enable unambiguous set-up. List minimum performance requirements and verification requirements, cross-referenced to the calibration section and any relevant instrument manuals. Number for later reference. For glassware, include grade where applicable (bear in mind that use of a particular grade may require justification and that proof of compliance may be required). Include environmental requirements (fume cupboards, etc.).

Diagrams and flow-charts may assist clarity.

9 Sampling and Samples

This section is not intended to include sample selection, which will probably feature in separate sample plan documentation. Include sufficient detail to describe how the test portion is obtained from the sample received by the laboratory. Include storage, conditioning and disposal details.

10 Calibration

Identify the critical parts of the analytical process. These will have to be controlled by careful operation and calibration. Cross-reference to the relevant sections above. Include calibration of equipment – what needs to be calibrated, how, with what and how often? Consider the appropriate traceability of calibrants and highlight any limitations associated with available calibrants.

11 Quality Control

Describe the quality control procedures required, frequency of QC checks during batch analysis, pass/fail criteria, action to take in the event of a failure. Cross-reference to the relevant sections above.

12 Procedure

Describe the analytical procedure, cross-referencing previous sections as appropriate, including numbered reagents, apparatus and instrumentation. Where parameters are expressed (time, temperature, etc.) which are critical to the procedure, cross-reference to the relevant part of the calibration section. Indicate at which point in the analytical procedure the QC and calibration procedures should be performed.

If the extraction and/or sample clean-up procedures are particularly complicated, a separate document for these procedures may be justified.

13 Calculation

Lay out the equations for calculating the results, ensuring that all terms are clearly defined and derived. Indicate requirements for checking and cross-reference to QC requirements.

14 Special Cases

Include any modifications to the method necessitated by the presence or absence of specific components in the samples to be analysed.

15 Reporting Procedures, Including Expression of Results

Indicate how results should be reported, including rounding of numbers, final units, uncertainty and level of confidence.

16 (Appendix on –) Method Validation

Depending on the volume of data in support of the validation, it may be appropriate to list it here or provide reference to a separate file.

17 (Appendix on –) Measurement Uncertainty

The major sources of uncertainty relating to the method should be identified. Those contributions not used in the final calculation, because they are considered insignificant, should be mentioned. The overall uncertainty should be listed, together with an explanation of how it was derived. A more detailed treatment may be in a cross-referenced file.

18 Bibliography

Include any references that give background information to the development and validation of the method.

Chapter 5
Making Measurements

Learning Objectives

- To appreciate the importance of careful work in making measurements.
- To understand the role of calibration and quality control in making analytical measurements.
- To understand the importance of using standards, reference materials and quality control samples.
- To recognize the characteristics of a laboratory environment which can affect the performance of equipment and hence influence the validity of measurements.
- To understand the principles for selecting equipment for particular applications.
- To be familiar with the need for equipment performance checks.
- To understand the importance of correct labelling of samples, chemicals and equipment.

5.1 Good Laboratory Practice

Good laboratory practice (or good scientific practice) is the term used to describe how chemists, and indeed other scientists, should go about their day-to-day work. It covers a variety of characteristics, such as tidiness, cleanliness, care, thoughtfulness, safety, organization and self-discipline. A chemist who has these qualities and uses them is more likely to get the correct results, and get them first time, than someone who does not have them.

Quality Assurance in Analytical Chemistry E. Prichard and V. Barwick

Note that good laboratory practice (glp) should not be confused with Good Laboratory Practice (GLP). The latter is the name given to a set of principles governing the organizational process and the condition under which non-clinical health and environmental safety studies are planned, performed, monitored, recorded, archived and reported, and was put forward initially by the Organization for Economic Co-operation and Development (OECD) (see Chapters 2 and 9).

Chemists whose work is characterized by good laboratory practice will work in such a way that they make sure they fully understand what it is they have to do before they begin work. They will always be in control of their actions.

5.1.1 Before Starting an Analysis

Suppose you are given the task of analysing a batch of samples using a particular, previously validated method. Clearly, you should not just rush headlong into the work, but should plan what needs to be done and when, and what equipment, reagents, etc., are required to carry out the work.

DQ 5.1

As a brief exercise, make a list of all of the points you need to consider before you would start work on the samples. Once you have made your list, have a look through the next section and see how your list compares.

Answer

Before starting work, the chemist should:

- Locate the samples.
- Ensure that an up-to-date copy of the method is available.
- Read the method if not already completely familiar with it.
- Check that all of the instrumentation required for the analysis is available for the period of the work.
- Check that all of the required instrumentation is in working order, clean, and, if appropriate, calibrated.
- Plan the sequence of the work, and what is required at each stage. Check whether any stages are critical and whether the method of analysis must be completed without any breaks or on the same day. The complexity of the method may, for instance, limit the number of samples that can be handled in a batch. Construct a simple timetable to help plan the work.

- Consider any hazards associated with the method and with the use of particular reagents.
- Consider any factors which may affect the results, such as past or present work, which may provide a source of contamination.
- Only start the work if an appropriate fume cupboard, fume hood, glove box or clean area is available. Hazards or contamination dictate where in the laboratory the work can be carried out.
- Allocate adequate clean bench space for doing the work, so that equipment can be laid out in an uncluttered way.
- Ensure appropriate safety clothing is available; usually, a laboratory coat and safety glasses are sufficient but certain methods will require additional equipment, such as gloves.
- Make other staff aware of potential problems. Arrange specialist first-aid requirements before starting the work. Where certain hazards are involved, supervision may be necessary.
- Check that any glassware required is clean, undamaged and, if appropriate, calibrated. Sufficient glassware should be collected before starting work. Note any particular precautions which may apply to cleaning glassware or other equipment, e.g. volumetric glassware should not be dried in an oven after cleaning as this can lead to permanent distortion and thus loss of calibration (see Section 5.6.5).
- Check reagents, standards and reference materials to ensure that adequate stocks of the correct grades are available. Where reagents and quality control samples require preparation, this may need to be done in advance. If stocks of prepared reagents already exist, these must be checked to ensure that they are still usable. All reagents should be well-labelled.
- Plan necessary disposal procedures, for example, for used samples, reagents, and contaminated equipment.
- Plan cleaning procedures for equipment.
- Check that evidence is available that you are competent to carry out the method.

Depending on your laboratory experience, you may not have noted all of the points but you may have considered other things not listed. The whole list may not apply to all methods. If in doubt, read through the list and consider why each point is there. The 'golden rule' is to be clear about what you are going to do before you start, and to have everything you need ready to use. Try to organize the work so that you have plenty of time to do each part without needing to hurry. Appreciate how long each part of the work will take and identify critical steps.

5.1.2 During the Analysis

Once work has begun the following list indicates what needs to be considered:

(i) for each sample, examine all of the details provided with the sample and note the sample condition – then cross-reference against associated paperwork;

(ii) check samples are at the correct temperature before opening their containers;

(iii) carry out sampling procedures, ensuring that each aliquot is sufficiently well-labelled at each stage of the analysis in order to be traceable back to the original sample;

(iv) where equipment is used several times for different samples, ensure adequate cleaning between each use to prevent cross-contamination;

(v) unless the method indicates otherwise, the correct sequence to follow is to carry out any necessary calibration; if the calibration is satisfactory, carry out the quality control (QC) checks; if the QC checks are satisfactory, carry out the sample analysis;

(vi) where samples are examined in batches, periodic checks on calibration and quality control may be necessary during the analysis of the batch;

(vii) follow the method exactly as it is written; do not be tempted to take short cuts – these will only lead to problems and inevitably prolong the analysis;

(viii) do not hurry the work – this results in mistakes; good planning reduces the need to rush;

(ix) record observations, data, and unusual method details clearly, in accordance with the recommendations given in Chapter 8.

The key point to pick up here is that good planning before starting the work, and working carefully and steadily, keeps problems to a minimum.

5.1.3 After the Analysis

After the practical work is completed, there are a number of things still to be done, as follows:

(i) Using the data gathered, calculate the required results, looking for obvious errors such as poorly matching values for duplicate samples, positive results where negative results were expected, etc.

(ii) Check data transcriptions and calculations. This is often better performed by someone other than the original person carrying out the work. The person doing the checking does not necessarily need to be senior to the analyst, but must understand the principles behind the work being checked. If you are in a group of analysts, you can check each other's work.

(iii) Samples should be retained at least until a satisfactory report has been produced. A sample may be retained for a further period of time, according to laboratory policy, or returned to the customer, or disposed of. Any sample disposal should be in accordance with laboratory safety rules (these should be formulated in compliance with national safety legislation (see also Chapter 4, Section 4.3.9)).

(iv) The laboratory area used for the work, any related equipment and instrumentation should be decontaminated, cleaned and tidied up ready for the next task. Reagents and chemical standards having a short shelf-life should be disposed of, with due regard to safety regulations.

The final points to pick up here are that care needs to be exercised at every stage of the work. Even then, simple errors may creep into the most carefully made measurements, but these can usually be spotted by cross-checking. Poorly matching results of duplicate analyses may well be a clue to problems in the whole measurement system. A key part of any work is to tidy everything away afterwards, and deal yourself with any hazards you may have created. In brief, leave everything as you would hope to find it.

SAQ 5.1

For each of the following statements, state whether you think it is true or false.

(a) If an analyst works carefully, there is never any need to make checks on results.

(b) A copy of the method should be available before starting work.

(c) Before starting work the analyst should assess the likely hazards, ensure that appropriate safety clothing is available and that other people working nearby are aware of the hazards.

(d) It is not the analyst's job to clear up after he/she has finished work.

(e) Samples can always be analysed straight from a refrigerator.

(*continued overleaf*)

SAQ 5.1 (*continued*)

(f) Volumetric glassware can be quickly rinsed and dried in a hot glassware oven before re-use.

(g) Satisfactory calibration and quality control checks are necessary before samples can be analysed and the data accepted.

(h) Short cuts are a quick and reliable way of speeding up sample analysis.

(i) If a method has been validated, it will always give the correct answer whoever uses it.

5.2 Calibration of Measurement

Making a measurement of any kind involves comparing the unknown (i.e. the test sample being measured) with a standard. The standard provides the link to the measurement scale being used (e.g. a ruler to measure length, a standard weight to measure mass, a pure chemical substance to determine the amount of a compound present). This is illustrated in Figure 5.1.

Calibration is an essential part of most measurement procedures. However, the terms 'calibration' and 'verification' are often used incorrectly, as if they had exactly the same meaning, but each has a quite distinct meaning. **Calibration** is a set of operations that establishes, under specified conditions, the relationship between an instrument response and the corresponding values of a standard, e.g. known concentration of a chemical substance or the mass value assigned to a check weight [1]. **Verification** is confirming by measurement that some specified requirement has been fulfilled, e.g. the balance is still behaving in accordance with the calibration certificate [2]. The difference is that calibration permits the estimation of errors of the measuring instrument or the assignment of values to marks on arbitrary scales, whereas verification of measuring equipment provides a means of checking that the deviations between values indicated by a

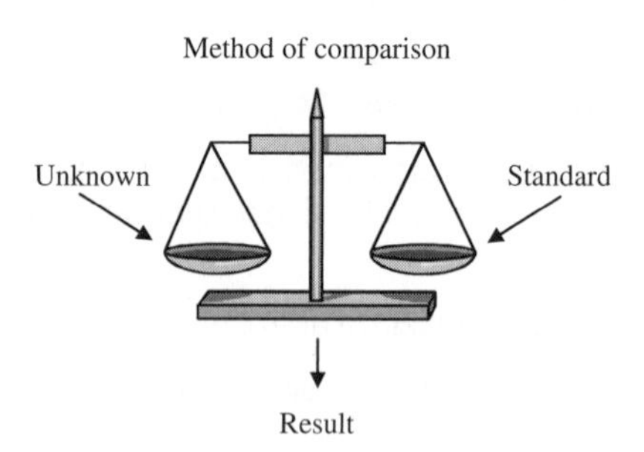

Figure 5.1 Illustration of the principles of measurement.

measuring instrument and corresponding known values of a measured quantity are acceptable. Using a 'check-weight' on a balance is verification that the balance gives a reading that is close enough to the known value to enable the analyst to use the equipment.

The term 'calibration' is often used when in fact what is meant is 'verification'. Calibration of an instrument or a piece of equipment (e.g. glassware) involves making a comparison of a measured quantity against a reference value. For example, to calibrate a spectrophotometer response, the appropriate reference material is selected and the spectrophotometer response to it, under the specified conditions, is measured. Then, the measured value is compared to the value quoted in the literature. Either a correction is made to the results from subsequent measurements or an adjustment is made to the instrument.

Calibration is also used to describe the process where several measurements are necessary to establish the relationship between response and concentration. From a set of results of the measurement response at a series of different concentrations, a calibration graph can be constructed (response versus concentration) and a calibration function established, i.e. the equation of the line or curve. The instrument response to an unknown quantity can then be measured and the prepared calibration graph used to determine the value of the unknown quantity. See Figure 5.2 for an example of a calibration graph and the linear equation that describes the relationship between response and concentration. For the line shown, $y = 53.22x + 0.286$ and the square of the correlation coefficient (r^2) is 0.9998.

Solutions of pure chemicals of known concentration used for instrument calibration are frequently referred to as 'standard solutions'. However, the term

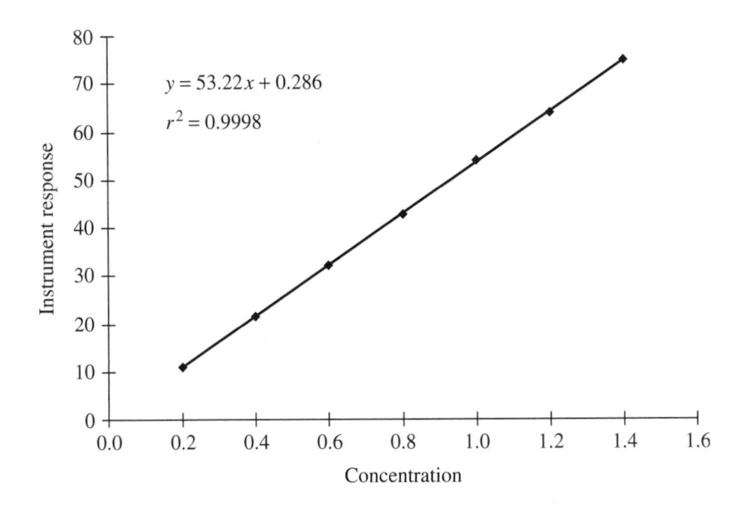

Figure 5.2 Illustration of a calibration graph.

'standard' has many different meanings; it may be interpreted in terms of a standard method (reference method) or a paper document such as the Standard, ISO/IEC 17025. Care should be taken in using this term to make sure that there is no ambiguity in the meaning.

Calibration using artefacts or materials with well-defined properties is used widely throughout the scientific community. In physical measurement, the fundamental base units (length, mass, time, electrical current, thermodynamic temperature and luminous intensity) are supported by a well-established system of internationally recognized standards, also known as *primary standards*. These standards are used to calibrate materials with less well-defined properties – these are known as *secondary standards, transfer standards* and *working standards*. The working standards are then used to calibrate the measuring instruments used to measure a particular property of the test item or determine the amount of the substance of interest in a test material. Since each standard has been compared to a standard higher up the chain, each with a specified uncertainty, it is possible to relate the result of the measurement directly back to the primary standard. This ability to relate back to a single standard is known *as metrological traceability* (see also Chapter 2, Section 2.1) [1]. Wherever possible, all measurements should show traceability to national or international standards. This is usually straightforward in physical measurement, but it is accepted that it can be problematical for chemical measurements.

Consider the following example. If a group of people was asked to measure the length of a line, each person using his or her own ruler, it is likely that each person would get a slightly different answer, even if they each closely followed the same method for making the measurement. The reason for this is due to the variation in the way rulers are made and graduated. Rulers might be made from wood, plastic or steel, each behaving differently as atmospheric conditions change. If, however, each person was able to compare the graduations on their ruler against that of a recognized 'standard' ruler, so that they could adjust their measured result by a correction factor, it is likely that the results would be much closer. This is a simple example of calibration. The 'standard' ruler provides a reference against which all the other rulers can be compared or calibrated. Each person's measurement of the line is therefore traceable to the 'standard' ruler. The use of calibration, and traceability to the 'standard' ruler, improves the *comparability* of the measurements.

Calibration using the appropriate physical standard in most cases is fairly straightforward. Taking a known mass of material is part of many analytical methods – calibrating the mass scale of a balance involves a physical calibrant such as a standard weight. The analyst has more of a problem with chemical calibration since there is no single standard that each measurand (e.g. concentration of cholesterol in serum) can be compared to. Metrological traceability in chemical analysis is achieved in two ways: first, by use of pure chemical substances, and secondly by using materials with typical matrices in which the amount of analyte present is well characterized. This latter type of standard is known as a *matrix reference material.*

The fundamental unit in chemical measurement is the *mole – amount of substance*. A mole is the amount of a substance that contains as many atoms, molecules, ions or other elementary units as the number of atoms in 0.012 kg of carbon 12 (^{12}C). It is the only dimensionless SI unit. In practical terms, it is almost impossible to isolate a mole of pure substance. Substances with a purity of better than 99.9% are rare; one exception is silver, which can be obtained with a purity of 99.9995% which is referred to as 'five nines silver'. Another problem is that it is not always possible to isolate all of the analyte from the sample matrix, and the performance of the chemical measurement may be matrix-dependent – a given response to a certain amount of a chemical in isolation may be different from the response to the same amount of the chemical when other chemicals are present. If it is possible to isolate quantitatively all of the analyte of interest from the accompanying sample matrix, then a pure chemical substance may be used for calibration. The extent to which the analyte can be recovered from the sample matrix will have been determined as part of the method validation process (see Chapter 4, Section 4.6.3).

To achieve traceability of a measurement result, everything that influences that result (the influence quantities) must be traceable to national or international standards [3, 4]. This is not too difficult to achieve in a laboratory that has a quality management system in place but requires a full understanding of the method being used. It also requires that the uncertainty associated with each influence quantity is known. One way of obtaining metrologically traceable results is to obtain the results by using a *primary method*. A primary method is defined by the International Bureau of Weights and Measures (BIPM) as a method having the highest metrological qualities, whose operation can be completely described and understood, for which a complete uncertainty statement can be written down in terms of SI units. Examples of primary methods include titrimetry and gravimetry. In many cases, it is not possible to use a primary method and other methods are acceptable. Two methods which have been used are isotope dilution mass spectrometry and neutron activation analysis. The general route to achieving traceability is shown in Figure 5.3.

What is essential in establishing traceability is that the measurand is specified unambiguously. This may be, e.g. in terms of extractable cadmium from soil by using a named acid mix or the concentration of a metal in a particular oxidation state, e.g. Fe(II) or Fe(III). The units used to report the result should also be known and acceptable; SI units are preferred. The method used will be validated and if used in accordance with the written procedures should produce results that are 'fit for purpose'. The class of glassware to be used will be specified in the method procedure, e.g. Class A pipettes and volumetric flasks, as these are manufactured to a specified tolerance. Instruments will be regularly calibrated and their performance verified daily. In terms of the chemicals used, these will

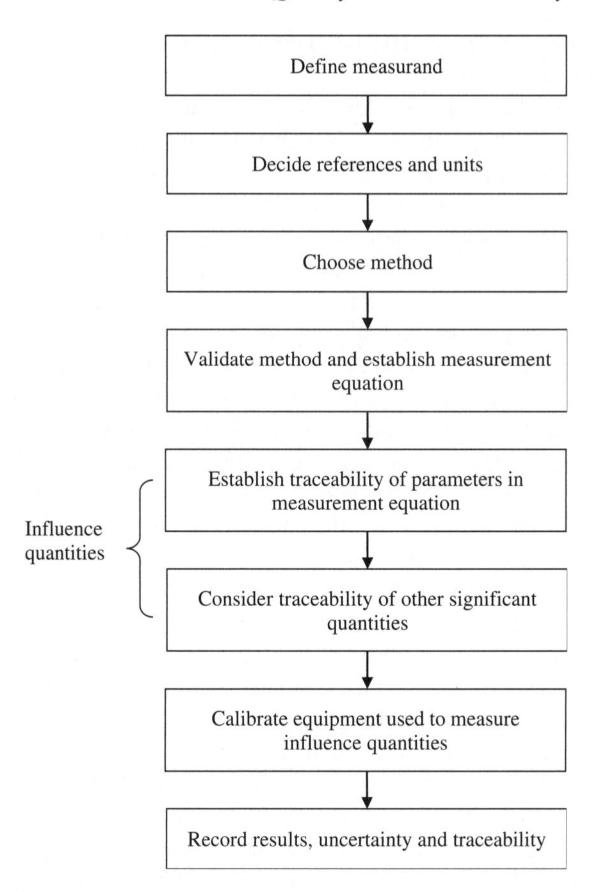

Figure 5.3 A strategy for achieving traceability.

be either substances of known purity or reference materials. Everything that appears in the measurement equation has to be traceable. In addition, there may be other variables which influence the result but do not appear in the measurement equation. Frequently, temperature, time and pH have to be controlled if consistent results are required. If this is the case, then these must also be traceable.

5.3 Achieving Metrological Traceability

Section 5.2 introduced the subject of metrological traceability and calibration and the use of pure chemical substances and reference materials in achieving traceability. Reference materials are used as *transfer standards*. Transfer standards are used when it is not possible to have access to national or international standards or primary methods. Transfer standards carry measurement values and can be used to calibrate measurement systems and validate methods. They can also be used to establish identity. Examples of transfer standards include reference materials (see below), physical standards (mass, temperature) and reference values (atomic values). **Reference material** is a generic term for a particular class of materials used as transfer standards in chemical measurement. There are several definitions; the most recent has been published in ISO Guide 35 [5], while a discussion of the definition can be found in Emons *et al.* [6].

5.3.1 Reference Materials

A **Reference Material** (RM) is defined as follows:

> *Material, sufficiently homogeneous and stable with respect to one or more specified properties, which has been established to be fit for its intended use in a measurement process.*

Notes added to this definition are:

(1) RM is a generic term.

(2) Properties can be quantitative or qualitative, e.g. identity of substances or species.

(3) Uses may include the calibration of a measurement system, assessment of a measurement procedure, assigning values to other materials and quality control.

(4) An RM can only be used for a single purpose in a given measurement.

Note (4) is very important as it highlights the fact that the reference material used for the method validation **cannot** be used again when the method is in routine use for calibration purposes. The same type of material can be used, but it needs to come from a different supplier. The same material cannot be used for calibration purposes and then as a quality control material.

Examples of RMs include the following.

Pure substance RMs: pesticides with quoted purities, polycyclic aromatic hydrocarbons with quoted purities and potassium hydrogen phthalate with a quoted purity.

Standard solutions: a solution of nickel in acid with a quoted mass/volume concentration, a solution of sodium hydroxide with a quoted concentration as a molarity and a solution of pesticides with quoted mass/volume concentrations.

Matrix RMs – natural materials: river sediment with quoted concentrations of metals, milk powder with a quoted fat content and crab paste with quoted concentrations of trace elements.

Matrix RMs – spiked materials: lake water fortified with trace elements and milk powder spiked with organic contaminants.

Physico-chemical standards: benzoic acid with a stated melting point, *p*-xylene with a stated flash point, sand with a quoted particle size distribution and polymers with quoted molecular weight distributions.

A **Certified Reference Material** (CRM) is defined as follows:

A reference material characterized by a metrologically valid procedure for one or more specified properties, accompanied by a certificate that states the value of the specified property, its associated uncertainty and a statement of metrological traceability.

Notes added to this definition are:

(1) The concept of value includes qualitative attributes, such as identity or sequence. Uncertainties for such attributes may be expressed as probabilities.
(2) Metrologically valid procedures for the production and certification of reference materials are given in, among others, ISO Guide 34 [7] and ISO Guide 35 [5].
(3) ISO Guide 31 gives guidance on the contents of certificates [8].

Some reference material producers/suppliers use different names to describe their materials. For example, a **Standard Reference Material** (SRM) is a certified reference material issued by the National Institute of Standards and Technology (NIST), while **European Reference Materials** (ERMs) are CRMs produced under a joint collaboration between three European reference materials producers, i.e. BAM (Federal Institute for Materials Research and Testing, Germany), IRMM (European Commission, Joint Research Centre, Institute for Reference Materials and Measurements, Belgium) and LGC (UK).

The term 'chemical standards' is used to describe chemicals (usually single chemicals) for which the purity is well-characterized.

Certified Reference Materials have many different uses, including the following:

- establishing metrological traceability of results;
- confirmation of the identity of a material;
- calibration and verification of measurement processes under routine conditions;
- verification of the correct application of standard methods;
- development and validation of new methods of measurement;
- defining values for other materials that may be used as secondary standards/calibrants;
- internal quality control and quality assurance schemes.

The development and characterization of Certified Reference Materials is an expensive process. Because of this, emphasis on the use of Certified Reference Materials is usually directed more towards the initial validation of a method; it is rarely economical to use a reference material for routine quality control although it can be used to 'calibrate' other, cheaper, secondary materials which can be used for routine quality control.

5.3.2 Chemical Standards

Chemical standards may be used for calibration in two different ways. They may be used 'externally', where they are measured in isolation from the samples, or 'internally', where the standard is added to the sample and both the standard and sample are measured at the same instant, i.e. as a single 'enriched' sample. These two approaches are frequently termed 'external standardization' and 'internal standardization', (see Sections 5.3.2.1 and 5.3.2.2 below).

5.3.2.1 External Calibration

This involves the use of one or more chemical standards made up to a known concentration and analysed at the same time as the samples as a means of quantifying the analyte concentrations in the samples. Pure chemical standards at a single concentration may be used to establish instrument response factors (the expected change in measurement signal per change in analyte concentration). During the method validation study, chemical standards at several concentrations (usually at least seven, including a blank) are used to prepare a calibration graph. If, during such a study, it has been demonstrated that the relationship between response and concentration is linear it is possible to use fewer concentrations for the calibration when the method is used to analyse samples. The response factor at a given concentration is therefore equal to the gradient of the graph at that point, i.e. the *sensitivity*. The analyte concentration in samples is estimated from the calibration graph by reading the concentration on the graph equivalent to the measured response. Note, when using these graphs, it should **not** be assumed that the graph passes through the point (0,0) unless measurement of a suitable blank has shown this to be the case. Calculation of sample concentrations using a calibration graph may **only** be performed in the concentration range between the upper concentration of the chemical standard and the lowest concentration of the chemical standard or blank value. Do not try to extrapolate at either end of the range. Remember that the predicted concentrations obtained from the calibration graph will be more certain at the mid-point of the graph and least certain at the graph's extremities.

The use of external chemical standards is suitable for many applications. Ideally, chemical standards should be matrix-matched with samples to ensure that they respond to the measurement process in the same way as the samples. In some cases, a sample preparation and measurement process has inherent faults which may cause losses of the analyte. In such situations, if possible, the chemical standards should be subjected to the same processes as the samples so that any losses occur in both samples and standards equally. In some situations, large errors occur during the introduction of the sample and standards into the instrument. An example is the use of manual injection of small volumes into a capillary gas chromatograph. It requires a very skilled technique to be able to inject the same volume repeatedly. A 10–50% variation is not uncommon. The influence of such a variation on the result can be eliminated by modifying the procedure. A chemical standard (A) is used to prepare the calibration graph and a second chemical standard (B) is added at the measurement stage to both the chemical standard solution (A) and to the sample solution. In this case, the ratio of the signal from A and from B is plotted against concentration. The analyte concentration is no longer dependent on injection size or technique.

5.3.2.2 Internal Standardization

Internal standardization involves adding a chemical standard to the sample solution so that standard and sample are effectively measured at the same time. Internal chemical standards can be either the actual analyte, an isotopically labelled analyte or a related substance. The last one is usually chosen as something expected to be absent from the sample yet expected to behave towards the measurement process in a way similar to the analyte. There are a number of different ways of using internal standards and they sometimes serve a different purpose.

As mentioned in the last section, when a related substance is added to both the chemical standards and to the samples problems with variations in injection volumes are removed. There is another use for internal standards of this kind, i.e. where the standard acts as an *internal calibrant*. The internal standard has to behave in the same way as the sample in relation to the measurement process, except that the signals can be distinguished from each other. When the related substance is added early on in the measurement process, any losses of analyte as a result of the measurement process are equally likely to affect the chemical standard and the analyte. Thus, no adjustment to the result, to compensate, e.g. for poor recovery, is necessary. The concentration of the sample is obtained from the ratio of the two signals (one from the standard and one from the sample).

Some methods may involve a procedure known as **standard additions**. This is when the internal chemical standard is identical to the analyte and a known amount of it is added to a sample solution. Clearly, if the internal chemical standard is the same chemical as the analyte, then in order to determine the analyte level in the sample, it will be necessary to measure the sample twice, i.e. once without any chemical standard added and once with the standard added. There are several ways of carrying out the process of standard additions; two are described here. The addition of a chemical standard which is the same as the analyte is also called ‘spiking’.

For the first standard additions method, only one portion of the standard is added to the sample. In this case, the original analyte concentration X is given by the following equation:

$$X = \frac{YAC}{[B - (DA)]} \tag{5.1}$$

where:

Y is the concentration of the added internal chemical standard;
A is the response of the unknown concentration of analyte;
B is the total response of the unknown concentration of analyte plus added chemical standard;
$C = (\text{Volume}_{\text{Standard}})/(\text{Volume}_{\text{Sample}} + \text{Volume}_{\text{Standard}})$;
$D = (\text{Volume}_{\text{Sample}})/(\text{Volume}_{\text{Sample}} + \text{Volume}_{\text{Standard}})$.

It is common to use internal chemical standards at high concentrations, added in small volumes with respect to the sample volume. In such cases, where the sample volume is much greater than the volume of chemical standard added, $D = 1$ and C becomes

$$(\text{Volume}_{\text{Standard}})/(\text{Volume}_{\text{Sample}}) = C'$$

with C' being effectively the dilution factor for the chemical standard. Equation (5.1) simplifies to Equation (5.2), as follows:

$$X = \frac{YAC'}{(B - A)} \tag{5.2}$$

An alternative standard additions procedure is to add several different known amounts of a chemical standard to the sample solution. Take equal volumes of the sample solution (seven portions are recommended). To all but one of these solutions, add different known amounts of pure analyte. Dilute all of the solutions to the same volume. Measure the response of each diluted solution. Plot a graph of response against concentration of analyte added. The negative intercept on the x-axis (at $y = 0$) represents the concentration of the analyte in the sample solution. This is illustrated in Figure 5.4, where the extended line from A to B gives a direct reading of the concentration in the sample. Since the graph of response against concentration is a straight line, this concentration can also be obtained by dividing the intercept by the slope. From the equation for the line shown in Figure 5.4 ($y = 69.1x + 7165$), the sample concentration is 103.7 μg l^{-1}. To minimize the uncertainty in the estimated concentration, it is important to use a large spread of concentrations (added amounts). Standard additions is a useful approach when external calibration is not possible because the response is affected by the sample matrix.

Neither external nor internal standardization techniques make allowance for the different behaviour of samples and chemical standards, due to the matrix effect in the samples or due to the different state of the analyte in the samples and

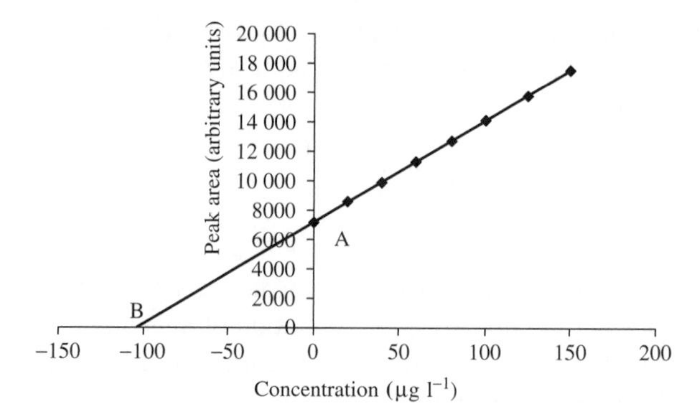

Figure 5.4 Illustration of a standard additions graph.

in the chemical standards. The chemical standard should be subject to the same analytical process as the samples, but even so, if the analyte is strongly bound in the sample matrix, it may not be possible to recover all of it in the analytical procedure. This should be borne in mind when designing analytical methods. Some form of detailed recovery study may be required to test how easily the analyte can be recovered from the matrix (see Chapter 4, Section 4.6.3).

Spiking a sample with a solution of the analyte of interest is also an effective way of confirming the presence of the analyte. In gas or liquid chromatography, it is not uncommon for the sample matrix to affect the chromatography, causing a difference in the retention times of the analyte peak in the sample and in the chemical standard. Thus, it is not certain whether the peak observed for the sample is due to the analyte or to some other artefact. By spiking the sample with the chemical standard and measuring the chromatographic retention time of the enhanced peak, it is possible to make this confirmation.

SAQ 5.2

(i) Use the following data to construct a calibration graph:

Chemical standard concentration (mg ml^{-1})	Instrument response (arbitrary units)
0 (blank)	5
2	10
6	18
8	22
14	35
20	47

(continued)

SAQ 5.2 (*continued*)

Determine the concentrations (to the nearest whole number) corresponding to samples having the following instrument responses:

(a) 15 (b) 36 (c) 55.

What further action is required for case (c)?

(ii) A sample is measured by using a method which involves an internal chemical standard. The chemical standard is identical to the analyte. The response for the analyte with no added chemical standard is 5. After 1 ml of standard (concentration = 5 mg ml^{-1}) is added to 9 ml of sample, the spiked sample is measured. The analyte response is now 15. Use this information to calculate the original concentration of analyte in the sample.

SAQ 5.3

The analysis of a soil for zinc has been carried out by the method of standard additions. The results obtained are tabulated below. Using a graphical method, estimate the concentration of zinc present in the sample solution.

Zinc added to the sample solution (ng ml^{-1})	0	10	20	30	40	50	60	70
Response (arbitrary units)	2.57	3.14	3.64	4.13	4.68	5.28	5.74	6.35

5.4 Quality Control

Quality control describes the measures used to ensure the quality of individual results or a batch of results. The measures used will vary according to the particular application. It is a means of evaluating the current performance of the method being used and the general procedures used in the laboratory. There are two types of quality control, namely internal quality control and external quality assessment. These are covered in detail in Chapters 6 and 7, respectively.

DQ 5.2

Suppose you are analysing a number of samples of ground nutrient tablets to determine their iodine content. The method is very simple. A known mass of each sample is taken, processed to isolate the iodine, which is then titrated against a sodium thiosulfate solution of known concentration. The end-point is determined by using a starch indicator. List the measures you could take which would help ensure the quality of your work and make your results more reliable.

Answer

Among the measures you might have considered taking are the following:

(i) Standardization of the thiosulfate solution against a recognized chemical standard so that its exact concentration is known.

(ii) Titration of reagent blanks – titration of the reagents used in the processing stage but without any sample present. This establishes whether anything other than the iodine present in the samples reacts with the sodium thiosulfate.

(iii) Titration of samples containing known amounts of iodine to determine whether 100% recoveries are being achieved.

(iv) Analysing samples in duplicate in order to check that the results are consistent.

What you actually carry out as quality control depends on the analytical problem you have to solve. The choice of actions are as follows:

- measure blanks;
- measure quality control samples;
- measure repeat samples;
- measure blind samples;
- measure chemical standards and spikes.

Before considering these in more detail it is worth considering why quality control is necessary. As mentioned above, it is used as a means of checking work to see whether the analytical system is working correctly. If the same measurement is made a number of times, we should not expect to get exactly the same answer each time. There will be a slight natural variation in the answers arising from small (and acceptable) variations in the analytical system (see Chapter 4, Section 4.6.2). Quality control is used to monitor whether the fluctuations observed in results are acceptable, due to the expected variation in the method, or whether they are due to some other unacceptable change. Repeated measurement of the quality control sample (Section 5.4.2) is the most commonly used monitoring system. The statistical theory behind quality control is more thoroughly dealt with in Chapter 6, where information about control charts can be found.

5.4.1 Blanks

Blanks are used as a means of establishing which part of a measurement result is not due to the characteristic being measured. The blank will be as close as possible in composition to the samples but *without* the analyte present. For example, you may dissolve a sample in nitric acid so that you can analyse it for traces of copper and nickel using atomic absorption spectrometry. As well as the sample containing traces of these metals, it is quite possible that the acid itself may contain traces of the same metals. A blank determination should therefore be performed by analysing the acid on its own – with no sample present. All of the reagents which may be used in processing the sample should be screened to ensure they do not affect the measurement. An analysis of all of the reagents with no sample present is known as a *reagent blank analysis*.

5.4.2 Quality Control Samples

Quality control (QC) samples are used as a means of studying the variation within and between batches of a particular analysis. A QC sample is a material that is fully characterized in-house or by a third-party, similar in composition to the types of samples normally examined, stable, homogeneous and available in large quantities so that it can be used over a long period of time, providing continuity. A reference material can be used as a QC sample. Its stability and homogeneity should ensure that variations in the results obtained during its analysis are due to variation in the analytical method and not in the composition of the sample. The variation in the result obtained from the quality control sample is normally monitored on a quality control chart (see Chapter 6, Section 6.2). These charts are used to demonstrate the statistical state of the measurement system – is it still in 'statistical control', showing signs of going out of control or already out of control? Results for test samples cannot be used if the method is out of statistical control and the reason must be investigated.

5.4.3 Repeat Samples

Repeat samples provide a less formal check than conventional QC samples. Within an analytical process, samples may be analysed singly, in duplicate, in triplicate, etc. Normally, the repeat sample is a conventional sample, repeated later in the batch of samples, or perhaps in a different batch. The variation between the two sets of results is studied to ensure that the variation is within the acceptable limits (see Chapter 4, Section 4.6.2). Higher than expected variation (for example, variation greater than the stated repeatability for the method) provides an indication that there is a possible fault in the analytical system. The analyst is normally aware when repeat samples are used.

5.4.4 Blind Samples

Blind samples are types of sample which are inserted into the analytical batch without the knowledge of the analyst – the analyst may be aware that blind samples are present but not know which they are. Blind samples may be sent by the customer as a check on the laboratory or by laboratory management as a check on a particular system. Results from blind samples are treated in the same way as repeat samples – the customer or laboratory manager examines the sets of results to determine whether the level of variation, between repeat measurements on the blind sample or between the observed results and an expected value, is acceptable, as described in Section 5.4.3.

5.4.5 Chemical Standards and Spikes

There are two uses of chemical standards in chemical analysis. In the first place, they may be used to verify that an instrument works correctly on a day-to-day basis – this is sometimes called **System Suitability checking**. This type of test does not usually relate to specific samples and is therefore strictly quality assurance rather than quality control. Secondly, the chemical standards are used to calibrate the response of an instrument. The standard may be measured separately from the samples (external standardization) or as part of the samples (internal standardization). This was dealt with in Section 5.3.2.

5.5 Environment

It is unfortunate that many analytical chemists are required to work in laboratories which are far from suitable for the type of tests that they are required to perform. This can ultimately influence the quality of the results they produce. There are a number of factors that may influence the quality of analytical work. One important factor is that when a sample is being analysed to detect, e.g. very small amounts of the analyte of interest, it is essential to avoid all other sources of the analyte and other potentially interfering species which might contaminate the sample and distort the result.

DQ 5.3

Before reading further, briefly list what you consider the sources of contamination to be.

Answer

The possible sources of contamination are:

- apparatus and equipment in contact with the sample;
- the analyst;

- other analysts;
- other samples;
- reagents and solvents;
- pure chemicals used to prepare standards;
- the laboratory atmosphere;
- the laboratory environment.

Contamination can be a very serious problem when carrying out trace analysis.

5.5.1 Factors Affecting Quality

The laboratory environment can affect the quality of measurements in ways other than direct contamination.

DQ 5.4

List the environmental factors that may influence a result.

Answer

You have probably identified factors such as vibration, dirt, sunlight, radiation, electric and magnetic fields and noise. Other factors, such as fluctuations in laboratory temperature and humidity, can have a more subtle effect.

Environmental factors are discussed in the following sections.

5.5.2 Laboratory Design

Where a laboratory is purpose-built, hopefully it will be designed in such a way as to minimize the environmental influences mentioned in Section 5.5.1. It can be very expensive to remove some of these problems and laboratory design often involves a trade-off between cost and reducing the effects of the problems. Where a laboratory is a conversion of an existing building, the trade-off can be even more severe. Many of these problems can at least be reduced by minor changes to the laboratory. The effects of sunlight can be reduced by fitting blinds, while filtered air conditioning can stabilize the temperature and humidity and reduce dust levels. Major sources of vibration may present problems, although these can be reduced by installing furniture designed to isolate delicate equipment from the effects of external sources of vibration.

5.5.3 Siting of Instruments

In setting up a new laboratory, the laboratory manager is faced with the task of taking the laboratory space, with all of its faults and imperfections and fitting in the various items of furniture and equipment in such a way that the equipment performs as well as possible.

For most chemists, the laboratory they work in will already be organized when they arrive. Everything will have been sited – hopefully in the correct place. This should not be taken for granted. Consistent poor performance of a particular piece of equipment should lead the analyst to question whether the environment might be the cause of the fault. The key message is to be aware which environmental factors will lead to poor performance in a particular method or in a piece of equipment.

5.5.4 Monitoring Changes

In a modern laboratory, automatic sensors are often used to detect unwanted changes in laboratory conditions and warn laboratory staff. Basic laboratory conditions, such as temperature, humidity and particulates, can all be monitored continuously using sensors. The results can either be fed to chart recorders, or into computer-controlled laboratory management systems, which can take corrective action or sound alarms in the event of the limit for a particular condition being exceeded.

5.6 Equipment and Glassware

In this section, we will consider equipment; this is everything other than chemicals that is used in a laboratory to enable analyses to be carried out. In the previous section, we saw how characteristics of equipment and the laboratory environment were interdependent, that care was needed, for instance, in siting equipment. In this section, the subject will be broadened and discussed in greater depth.

5.6.1 Selection

There are a number of factors which contribute to the choice of a particular piece of equipment for a particular application.

DQ 5.5

Consider the following two situations:

(a) choosing from existing equipment in the laboratory for a particular task;

(b) buying a new piece of equipment.

For each of these two situations, list as many things as possible that you would consider when choosing equipment.

Answer
When choosing from existing equipment, the things you should consider are suitability, i.e. 'fitness for purpose', condition, cleanliness and, in some cases, if its performance is up to the published specification. When buying new equipment, it is important to identify all possible applications of the equipment. Once this is done, a specification can be written, bearing in mind 'fitness for purpose', cost (initial and running costs) and ease of use. Less obvious points which may be important and relevant are, size, weight, power requirements, manufacturer's reputation for reliability, ease of servicing and availability of spares.

All equipment has limitations, for example, the amount of a substance it can detect or the accuracy with which it can make a measurement. If you attempt to make the equipment perform beyond its capabilities, it does not matter how carefully the equipment is operated, it will not be possible to get meaningful results. In terms of a particular instrument, 'fitness for purpose' is interpreted as having appropriate performance capability to do the work required. This applies to all equipment, large or small. For example, a stirrer needs to perform the intended task satisfactorily while remaining essentially inert. There is a formal process for assessing the suitability of equipment to perform a given task – this is called *Equipment Qualification* or *Equipment Validation*. This is dealt with in Section 5.6.3.

DQ 5.6

Suppose you need to select a stirrer to help mix some crystals in a concentrated mineral acid. Which of the following materials could be used for the stirrer so that it would not react with the acid?

(a) polytetrafluoroethylene (PTFE) – Teflon;

(b) nickel/chromium;

(c) iron;

(d) glass.

Answer
Iron is definitely unsuitable – the preferred choice would be PTFE. Glass or nickel/chromium might be suitable, depending on the type of mineral acid used and the analysis to be undertaken.

It is worth noting from the above discussions that, when selecting equipment, it is assumed that equipment received as new will be in working order, but it is also assumed that it will break down sooner or later! The reputation of individual manufacturers is worth considering when selecting equipment, both in terms of the actual products and the quality of their back-up services. The equipment which costs least to begin with is not necessarily the one which costs least when in use over a long period of time. For equipment in use, consideration centres on the fact that the items of equipment may not have been used as carefully as they should. You need to verify that they have not deteriorated to the extent that they are unsuitable for the application.

Physical details of equipment, such as size and weight, are also important. Suppose you occupy a laboratory on the first floor of a building, the only access to which is via a tight stairwell. You purchase an item of major equipment without noting its size and weight. Imagine the problems when it arrives and it is not possible to get it through the laboratory entrance not to mention up the stairs. In any case, if you could get it up into the laboratory, the floor might collapse under the weight.

5.6.2 Suitability

Before using any item of equipment, its condition must be verified as suitable for the application. If it is suitable, no problem; if not, what is needed to restore it to a satisfactory condition? Are there applications for which it would be suited in its current or partly restored state? If the answer is no to all of these, then there is little point keeping the item, and disposal will, at least, ensure it is not used by accident. Where something can be repaired but perhaps not immediately, it must be labelled as such, e.g. 'defective and awaiting repair', so that it is not used by accident.

Glassware is a special case. It is particularly vulnerable to damage and only in the case of expensive intricate items is repair ever viable. The usual course of action is to discard any damaged glassware. Even minor damage, such as chips, can result in subsequent failure which could be both costly and dangerous. Using damaged glassware is rarely worth the risk. Volumetric glassware should be disposed of however minor the damage, since repair is likely to adversely affect the volumetric characteristics of the item, such as graduated volume, rate of delivery, etc. Calibration of glassware (as well as cleaning) is important if, e.g. the glassware volume directly influences the results (calibration solutions, titrations, etc.).

5.6.3 Equipment Qualification

The primary requirement for all equipment (whether it be a volumetric flask, an oven used for drying samples or an atomic absorption spectrometer used for determining trace metal concentrations) is that it must be fit for its intended purpose. Other laboratory activities, e.g. method validation, assume and depend on the correct functioning and 'fitness for purpose' of any equipment used. GLP and International Standards, such as ISO 9001 and ISO/IEC 17025, require that instruments are suitable for their intended use. In addition, regulatory authorities and accreditation bodies are increasingly seeking evidence that equipment is 'fit for purpose'. This is what formal equipment qualification provides. Equipment Qualification (Equipment Validation) is a formal process that provides documented evidence that an instrument is fit for its intended purpose and kept in a state of maintenance and calibration consistent with its use [9]. It is the formal nature and need for documented evidence that makes this slightly more rigorous than what is done routinely in most laboratories.

The process of qualification involves four phases:

- Design qualification (DQ);
- Installation qualification (IQ);
- Operational qualification (OQ);
- Performance qualification (PQ).

Design qualification is the planning stage before a new instrument is purchased or before an existing instrument is selected for use for a particular task. Consideration has to be given to what the instrument will be required to do and what represents 'fitness for purpose'. The extent to which this is done will depend on the complexity of the task and the risk of making a wrong decision. Deciding whether an existing instrument is suitable for use on a new method might be straightforward. It might involve no more than a paper check to ensure that the instrument is, in theory, capable of operating over the range of concentration required, and at a level of bias and precision required by the method. Defining 'fitness for purpose' for a new instrument might be a complex exercise culminating in the development of a formal specification that states the required performance characteristics of the instrument and other factors relating to its use.

Installation qualification is aimed primarily at new instruments. This is the stage when the checks are carried out to confirm that the instrument received is as specified and correctly installed in the selected environment. This includes both hardware and software. It may be convenient to use a 'check-list approach' to this phase as that ensures everything is checked. This stage covers the installation up to and including its initial response to power, if that is relevant. In addition it may be appropriate to repeat aspects of IQ following relocation or upgrades of instruments.

Operational qualification establishes that an instrument will function according to its specification in the selected environment. The role of OQ can be considered as demonstrating that an instrument's key operating parameters are within specification and there are no unacceptable differences between the parameters selected and actual values. For example, when a pump is set to deliver 1.0 ml min^{-1} the actual flow is within the required tolerance (0.95 to 1.00 ml min^{-1}) and it is not significantly different, e.g. 0.7 or 1.3 ml min^{-1}.

Performance qualification fulfils two purposes. Initially, it serves to demonstrate that the entire instrument functions correctly. Subsequently, it is used to show that although the instrument performance is changing (because of, e.g. wear to a pump piston seal) the current performance is still 'fit for purpose'. Evidence that the entire instrument is functioning correctly can be obtained by using either an independent system test (e.g. in chromatography employing a 'test column' and 'test mix') or from everyday method-related checks (system suitability checking, calibration and analytical quality control). The use of a test column and a test mix has advantages. It enables the performance of the instrument to be evaluated by using a well-characterised test procedure and compared with that obtained previously or in the future. It also enables an instrument's performance to be compared with that of other instruments, either in the same laboratory or elsewhere.

This may appear to require a great deal of extra effort. In fact, this is not the case when you consider when these checks are made. The DQ stage is done once, either when the instrument is being selected from existing equipment or when a new instrument is being purchased. The IQ phase is mainly a one-off check, when the instrument is delivered or moved. The OQ phase is carried out regularly but not frequently. OQ should always be performed after initial installation of a new instrument. The frequency of future OQ testing depends on a variety of factors. These include the following:

- 'criticality' of the performance of the instrument;
- the level of use of the instrument – high workloads may cause the instrument to deteriorate more quickly and require thorough performance testing before use;
- the nature of the use of the instrument, e.g. use of aggressive solvents causes faster deterioration of pump components;
- the environment – an instrument in a mobile laboratory may require more frequent testing than one located in an air-conditioned laboratory;
- manufacturer's recommendation.

It is always recommended that OQ is carried out after a service visit. Evidence of continued satisfactory performance during use (PQ) should be obtained from everyday method-related checks (e.g. system suitability testing, calibration and analytical quality control). It is advisable to set up thresholds outside of which the performance of an instrument is no longer acceptable. Each stage of the equipment qualification needs to be fully documented so that the evidence of performance at any given time can be checked.

5.6.4 Cleaning

Cleaning is a form of maintenance which is particularly relevant where a piece of equipment is used repeatedly, but is also applicable to decontamination of equipment after use in 'dirty' environments. The purpose of cleaning is to ensure that when the piece of equipment is used for an application or measurement, the risk of contamination from previous samples, chemicals, standards or the laboratory environment will be minimized. In the majority of cases, the process of cleaning introduces new chemicals to whatever is being cleaned. After cleaning, the equipment must be well rinsed to remove all traces of the cleaning chemicals, and then dried.

Care should therefore be taken during cleaning to ensure that the cleaning process does not cause more problems than the original contamination. Some of the potential problems are fairly obvious, while others are more subtle, particularly where chemical reactions can cause physical changes to the equipment. For example, the operation of U-tube viscometers depends on the principle of capillary action, which in turn is dependent on 'wettability' and surface tension. Cleaning with certain solvents or detergents/surfactants may cause irreversible changes to the wettability of the capillary surface. When cleaning volumetric glassware avoid, if possible, machine washing, extremes of temperature and the use of strong acids/alkalis or surfactants.

The cleaning of delicate instruments or their components may cause damage which results in worse problems than those caused by the original contamination. Unless clear cleaning instructions, aimed at the analyst, are given in the instrument manual, it is safer to leave cleaning to the maintenance engineer. If in doubt, leave it to the expert. Other sources of guidance may be available from cleaning agent manufacturers. Where cleaning procedures are established for important applications, they should be documented.

5.6.5 Drying

Care should also be taken when drying equipment, after it has been cleaned. There are essentially four ways of drying equipment i.e. physically drying the item with an absorbent material, rinsing with a volatile solvent and allowing this solvent to evaporate at room temperature, driving off solvents at room temperature by using an air stream, or finally by driving off any solvents at elevated temperatures. The last approach is convenient and usually safe. Most laboratories have a commercially produced glassware drying oven. The temptation is to dry all glassware this way; however, there are two particular applications where it is unsuitable. A hot oven or any other source of heat is not recommended as a means of driving off volatile organic solvents. Similarly, heat should not be used to dry volumetric glassware. Glass expands when heated and the expansion may not be completely reversible. Thus, heating must be avoided since it may cause volumetric glassware to fall outside of its calibrated tolerances. A more suitable means of drying is to use an air stream; in the case of organic solvents the immediate area should be well-ventilated or preferably the drying carried out in a fume cupboard.

5.7 Chemicals and Consumables

This section deals with the correct use of chemicals and other consumable materials used in the course of chemical analysis. It includes advice on solvents, reagents (substances which play a specific role or have a specific reaction as part of a chemical test) and other materials which are used in chemical tests but do not take part in chemical reactions. A few examples of materials used are given below to illustrate this. Reference materials play a special role and are dealt with above in Section 5.3.

Reagents – reducing and oxidizing agents, indicators, drying agents, buffer solutions, complexing agents, acidic and alkaline materials.
Solvents – water, organic liquids and supercritical fluids.
Consumables – filter papers, anti-bumping granules, Soxhlet thimbles and column packings for chromatography.

For each of these you need to consider a number of aspects, such as grade, labelling, preparation, containment, storage, safety, stability and disposal. Each one of these is considered in the next section. Much of the advice given in this section is also applicable to samples.

5.7.1 Grade

Most laboratory chemicals are available in a number of grades, usually according to the concentration of impurities that are present. Generally, the purer the chemical, the more expensive it will be. A supplier's catalogue will indicate the different grades available for a particular chemical, together with the related purity specifications. You should bear in mind that the specification may not identify all of the impurities that are present. The nature of the impurities may or may not be important, depending on how the chemical is to be used.

For example, the industrial preparation of mineral acids, such as sulfuric, hydrochloric and nitric, inevitably leads to them containing small concentrations of metals as impurities. If the acid is to be used purely as an acid in a simple reaction, the presence of small amounts of metals is probably unimportant. If, however, the acid is to be used to digest a sample for the determination of trace metals by atomic absorption spectrometry, then clearly the presence of metallic impurities in the acid may have a significant effect on the results. For this latter application, high-purity acids that are essentially metal-free are required.

Similarly, organic liquids have a variety of applications. For example, hexane, which frequently contains impurities such as aromatic compounds, is used in a variety of applications for extracting non-polar chemicals from samples. The presence of impurities in the hexane may or may not be important for such applications. If, however, the hexane is to be used as a solvent for ultraviolet spectroscopy or for HPLC analysis with UV absorbance or fluorescence detection, the presence of aromatic impurities will render the hexane less transparent in the UV region. It is important to select the appropriate grade for the task you have. As an example, three different specifications for *n*-hexane ('Distol F', 'Certified HPLC' and 'Certified AR'), available from Fisher Scientific UK, are shown in Figure 5.5 [10]. You will see that the suppliers provide extra, valuable information in their catalogue.

Sometimes, extra chemicals are added to the main chemical as stabilizers. For example, formaldehyde is too reactive in its pure state to exist as formaldehyde for any length of time. It will dimerize or polymerize on standing. Formaldehyde is normally sold as a 40 vol% solution in water, with a methanol stabilizer (12 vol%) to slow down the polymerization.

SAQ 5.4

You have the task of purchasing some *n*-hexane for use in three different applications: (i) pesticide analysis by gas chromatography, (ii) as a solvent to extract some non-polar high-boiling (200–300°C) oils from a soil sample, and (iii) as a mobile phase for HPLC analysis with UV detection. List and contrast the performance characteristics you need to take into account for purchasing the appropriate grade of hexane in each case. *n*-Hexane boils at about 70°C. Will any of your choices of hexane be suitable for use for HPLC analysis with fluorescence detection? Explain your decision.

5.7.2 Labelling

Labelling is a very important feature of laboratory management. Properly designed and used labels ensure that the identity and status of reagents, chemical standards, apparatus and equipment are always clear to users. There are various requirements that the label on a container should satisfy:

- it should be securely attached to the body of the container, NOT the closure;
- it should have sufficient space to record all relevant information;
- it should be sufficiently indelible or protected to prevent the information becoming illegible due to spillage or soiling.

There will be exceptions to the above, which is why it is important to consider the purpose of the label. When preparing solutions by weight, one often has a number of identical 'lidded' containers. In this case, it is important to identify

(a)

Hexane (95% *n*-Hexane approx)

Distol F

Application - **For residue analysis**

H/0403/15 - 1 L
H/0403/17 - 2.5 L

Residue after evaporation	<2 ppm
Water (Karl Fischer)	<0.01 %

GC Analysis

Pesticide Analysis

No impurity peak that has a retention time greater than Lindane will have a peak height greater than that produced by 10 pg/mL Heptachlorepoxide (ECD) and 5 pg/mL Parathion (NPD).

Hydrocarbon analysis

No impurity, whose chain length is greater than Dodecane, is present at a concentration greater than 0.01 ppm (by FID).

Conditions:

Column:	J&W DB5: 30 m × 0.32 mm × 0.1 μm df
Temp Prog:	1; 50°C, 2 mins
	2; 40°C/min up to 320°C
	3; 20°C for 30 mins
Injection:	5 μL cold on-column

(b)

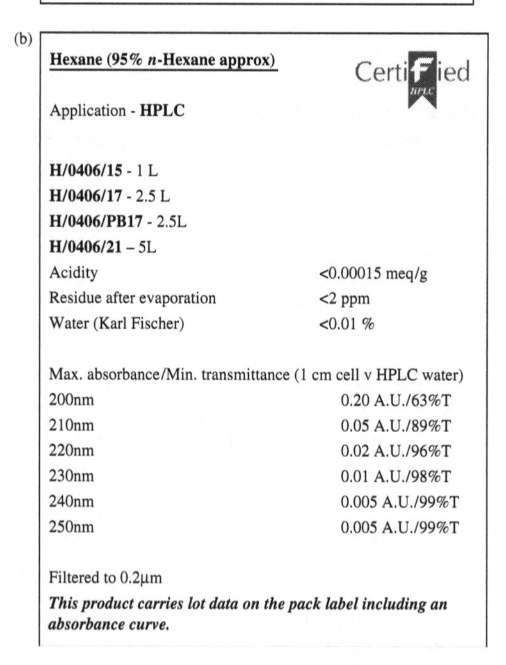

Hexane (95% *n*-Hexane approx)

CertiFied HPLC

Application - **HPLC**

H/0406/15 - 1 L
H/0406/17 - 2.5 L
H/0406/PB17 - 2.5L
H/0406/21 – 5L

Acidity	<0.00015 meq/g
Residue after evaporation	<2 ppm
Water (Karl Fischer)	<0.01 %

Max. absorbance/Min. transmittance (1 cm cell v HPLC water)

200nm	0.20 A.U./63%T
210nm	0.05 A.U./89%T
220nm	0.02 A.U./96%T
230nm	0.01 A.U./98%T
240nm	0.005 A.U./99%T
250nm	0.005 A.U./99%T

Filtered to 0.2μm

This product carries lot data on the pack label including an absorbance curve.

Figure 5.5 Examples of specifications for three grades of *n*-hexane: (a) Fisher 'Distol F'; (b) Fisher 'Certified HPLC'; (c) Fisher 'Certified AR'. Reproduced by permission of Thermo Fisher Scientific Ltd.

(c)

Hexane (95% *n*-Hexane approx)

Certified AR

Application – **For analysis**

H/0355/17 – 2.5 L

H/0355/21 – 5L

Assay (GLC)	>95.0%
Acidity/Alkalinity	<0.00008 meq/g
Aromatic hydrocarbons	<0.1 %
Bromine number	<0.5
Colour	<10 APHA
Residue after evaporation	<0.002%
Substance darkened by H_2SO_4	Passes test
Water (Karl Fischer)	<0.02 %
Calcium (Ca)	<0.2 ppm
Copper (Cu)	<0.02 ppm
Iron (Fe)	<0.1 ppm
Lead (Pb)	<0.02 ppm
Magnesium (Mg)	<0.05 ppm
Potassium (K)	<0.1 ppm
Sodium (Na)	<0.1 ppm
Zinc (Zn)	<0.1 ppm
Total phosphorus (P)	<0.1 ppm
Total silicon (Si)	<0.05 ppm
Total sulfur (S)	<0.001%
Typical analysis	
Aluminium (Al)	<0.1 ppm
Barium (Ba)	<0.02 ppm
Cadmium (Cd)	<0.02 ppm
Chromium (Cr)	<0.02 ppm
Cobalt (Co)	<0.02 ppm
Lithium (Li)	<0.02 ppm
Manganese (Mn)	<0.02 ppm
Molybdenum (Mo)	<0.02 ppm
Nickel (Ni)	<0.02 ppm
Strontium (Sr)	<0.02 ppm
Titanium (Ti)	<0.02 ppm
Vanadium (V)	<0.02 ppm

This product carries lot data on the pack label.

Figure 5.5 (*continued*).

the container and its cap. Some labels may not retain constant weight – in such cases, writing directly on the container may be the best option. This means that great care has to be taken when pouring liquids/solvents, to avoid obscuring the information on the label.

DQ 5.7

Before reading the section below, write down ten examples of where labels might be used in a laboratory and for what purpose. To help you, here is an example: labels used on individual instruments, each showing a unique number, to distinguish one from another of the same kind, and to enable their identification on the laboratory equipment inventory.

Answer
How did you get on? You have probably found that there are a virtually unlimited number of uses for labels in a laboratory. What you actually put on a label of course depends on the purpose of the label. The 'golden rule' is that the information on the label should be *clear*. This incidentally means that the label must be robust to exposure to sunlight or chemical spillages.

Chemicals and consumables normally arrive from the supplier in a suitable container, appropriately labelled. The information given on the packaging is the responsibility of the supplier and is legally required to conform to minimum requirements under packaging and labelling regulations. Typically, the label on a container for a commercially sold chemical will indicate the following:

- manufacturer's details;
- identity of contents (with alternative names), chemical formula and relative molecular mass;
- net weight or volume;
- grade (and percent purity);
- batch number;
- expiry date;
- special storage conditions, including temperature, humidity and light sensitivity;
- hazards and disposal instructions (according to recognized symbols and codes).

The label may also show other information:

- special use;
- detailed breakdown of impurities and their concentrations.

SAQ 5.5

Design labels for the following three applications:

(i) to inform that a piece of equipment is defective and must not be used;

(ii) to identify a volumetric solution for use for a specific application;

(iii) to identify a steel drum for use for waste solvents.

5.7.3 Preparation

It is often necessary to carry out some reagent preparation. This may seem a trivial aspect of laboratory work, but its importance is often underestimated. It is a common source of error and it is worth taking a little bit of time to ensure reagents and, in particular, standards are correctly made-up. Very simple principles are involved. Follow any instructions available, implement safety instructions, use equipment properly and check that you know what you are doing before you start.

Some instructions can easily be misinterpreted if not read carefully.

DQ 5.8

Consider the subtle differences in the following either/or statements.

Take 5 ml of ethanol in a volumetric flask and either (i) add 100 ml of water, or (ii) make up to 100 ml with water.

In each case, what would you expect the final volumes to be?

Answer

In (i), you would expect to end up with 105 ml and in (ii) you would end up with 100 ml. It may seem a very obvious point to make but it is typical of the small details which can be misread when using a method and is clearly a potential source of error. Similarly, weighing instructions appear in different forms.

DQ 5.9

Consider the following three weighing instructions:

(i) weigh about 1 g of sodium chloride;

(ii) accurately weigh about 1 g of sodium chloride;

(iii) take exactly 1 g of sodium chloride.

Decide which of the following five mass measurements satisfies the requirements for each of the above instructions:

0.9976 g, 1.1073 g, 1 g, 0.9 g or 0.9000 g?

Answer

All of the results satisfy (i).
0.9976 g meets the requirements of (ii).
1.1073 g meets the requirements of (ii).
0.9000 g meets the requirements of (ii).
Neither 1 g nor 0.9 g meets the requirements of (ii) and (iii). None of the results could be described as 'exactly 1 g'.

Subtle differences in the instructions can make a big difference in the way that a task is carried out and can have a significant effect on events if the instructions are not followed correctly. This works both ways. Obviously, if you weigh something approximately when it should be weighed accurately then you run into problems, but on the other hand if you weigh accurately when an approximate result is appropriate, you end up wasting time and effort.

5.7.4 Manipulation

The manipulation of chemicals, reagents and samples is an area where great care is needed in order to prevent contamination. For example, you may be sitting at a balance weighing out various chemicals into flasks for preparing reagent solutions, but using only one spatula. Clearly, the spatula may cause contamination if it is not cleaned thoroughly before moving on to the next chemical. Similarly, you should **never** put a spatula or pipette directly into the original reagent or solvent bottle, since this can cause contamination, however minor. The approximate amount of stock reagent or solvent should be poured from the main bottle into a clean beaker. The required measured amount can then be taken from the beaker by spatula or pipette without fear of contaminating the main stock. Anything remaining in the beaker should, **on no account**, be returned to the main stock as this will cause contamination. This principle should be applied to all situations where it is important not to contaminate the main stock. It should also be applied to situations where something is successively diluted; care should be taken not to contaminate back up the chain of calibration standards. Spatulas, pipettes, weighing boats and other equipment, which may be used a number of times for different applications, should always be treated as possible sources of contamination and cleaned between each use.

5.7.5 Containers

These come in a variety of forms, such as bottles, jars, cans and cylinders (for gases), and may be made from a number of materials. Containers are used in all aspects of analytical measurement, from the point of sampling (for samples) or production (for other chemicals or consumables) through the measurement, to disposal of the sample or reagents. At all points in the analytical process, it is vitally important that the container remains essentially inert with respect to its contents. A container must protect the contents from outside contamination while ensuring the contents are restrained from affecting the environment outside the container (see also Chapter 3, Section 3.6).

A container essentially has three characteristic features which may be either separable or inseparable, depending on the container. These three features are the container itself, the label (usually and preferably attached to the container) and the closure. A stoppered jar with an attached adhesive label is an example of a container in which the three features are *separable*. A polyethylene bag with an opaque label area and a 'press-together' closure is an example of a container in which the features are *inseparable*.

The choice of container is very much a matter of common sense. Reputable manufacturers generally supply chemicals or consumables in appropriate containers. If the contents are transferred to a different container, then this must be selected with care. In the case of sampling, obviously the person taking samples needs to have an appreciation of why the samples are being taken, how such samples should be stored to prevent spoilage and what safeguards need to be taken to ensure the samples are not spoiled. The chosen container and closure should be clean and inert with respect to the sample. The closure should seal the contents securely in the container.

The closure must effectively and safely seal the container, while remaining inert with respect to the contents. Inertness is often achieved by using a polytetrafluoroethylene (PTFE) washer on the inside of the closure. In some applications, some form of additional seal, 'double-containment' may be used because the contents of the container are either dangerous or proof is required that the container has not been tampered with and/or the contents disturbed (for example, forensic samples).

'Class A' volumetric flasks and their stoppers will normally have an identification mark so that they can be identified easily as the ones used in a particular analysis. When preparing solutions of known concentration a glass stopper is preferred.

SAQ 5.6

You need to take samples of (a) water suspected of pollution by organic compounds, (b) an unknown white powder and (c) diesel fuel containing anti-theft marker dyes. In each case, decide which of the containers in the following list would be suitable. You can use the same container for more than one application. Containers: (i) polyethylene bag with 'freezer-tie' closure, (ii) can with screw top, (iii) glass bottle and (iv) polypropylene bottle.

5.7.6 Storage

Once a sample has been taken it is important that it is properly stored until it can be analysed. Similarly, chemicals and consumables should be stored so as to preserve their condition and integrity. Storage conditions must be such that the chemical or sample does not undergo changes during storage and is unable to harm or otherwise affect its surroundings. This means that it is first contained within an appropriate sealed container, clearly and unambiguously labelled. Within this container, there may also be a preservative to help prevent deterioration of the contents. The type of container used will be dictated by the properties of the contents, as discussed in Section 5.7.5. The sample containers themselves will then be stored in a cupboard, storeroom, refrigerator, freezer or cold-room as appropriate (see Chapter 3). The choice will be dictated by the properties of the sample, typically protecting the contents from light, elevated temperature, air, humidity and dirt, as well as chemical, animal and microbiological attack.

It is common practice to make up chemical standard solutions in volumetric flasks and then store the solutions in those same flasks. Since volumetric flasks are both relatively expensive and fragile, such practice is not to be encouraged. Ideally, the prepared solution should be transferred to a suitable storage bottle. Before transferring, check that everything is dissolved, i.e. that the solution is homogeneous.

Where items are stored in close proximity to one another, such as in refrigerators, freezers and cupboards, there may be a significant risk of cross-contamination. Samples and concentrated solutions of chemical standards should certainly not be stored in the same storage area.

5.7.7 Safety

Safety is important in any chemical laboratory but is not normally considered a formal part of quality assurance procedures, unless the lack of safety also imperils the quality of the work. The relevance of safety is based on it being part of good operating practice within a laboratory and this in turn needs to be optimized in order to produce good quality results. Many of the chemicals used, and some of the samples encountered in a laboratory, are dangerous and certain rules should be followed to ensure that they can be handled safely. Most countries have a list of substances which have to be controlled carefully and the maximum level to which workers can be exposed.

The EU issued a Directive in 1998 covering the protection of health and safety of workers from the risks related to chemical substances [11]. In the UK there is a legal requirement based on this Directive, namely Statutory Instrument 2002/2677 [12]. The UK Health and Safety Executive issues guidance on implementing COSHH (Control of Substances Hazardous to Health) Regulations. Each laboratory is required to assess the risk associated with each chemical (or generic families of chemicals) in use in that laboratory. This risk is assessed according to the intended use of the chemical and the particular hazards associated with the chemical. Information on the latter is usually available from the manufacturer. Many European countries have similar legislation in place. Countries outside Europe have a broadly similar approach to the one adopted in Europe, i.e. risk assessment is paramount and very little is absolute in requirement.

Where a chemical is bought directly from a manufacturer, the analyst can refer to the label on the container for information on the various hazards presented by the contents. The manufacturer is nowadays obliged to provide hazard information on the label. However, this has not always been the case and there are many bottles of reagents in use in laboratories where the labelling is less than perfect.

Assuming that the analyst is not able to get the required information from the label, where else can it be found? **Hazard data sheets** are usually available for each chemical from the manufacturer. Likewise, the information is likely to be listed in manufacturers' catalogues. Failing this, there are various books which list chemical properties, including *The Merck Index* [13]. If all else fails, the chemist should assume the worst and treat the chemical with extreme caution.

Where a reagent has been prepared in the laboratory, and it is no longer in its original container, the chemist carrying out the preparation should ensure that the label carries any relevant information on hazards associated with the reagent itself and any solvents used to dissolve the reagent.

Very often the complete history of samples received into the laboratory may be unclear or incomplete. If it is not possible to find out detailed information about the background of a sample, then the sample should be treated with extra care. Where subsamples of a particular sample are sent to different parts of a laboratory, the labelling of each subsample should include any relevant safety warnings.

5.7.8 Disposal

Responsible disposal of chemicals, samples and consumables is likewise an important aspect of good operating procedures in the laboratory. Regulations are fairly strict in terms of what may be disposed of into the drainage system. It may be permissible to dispose of some chemicals directly down the drain, flushed down with copious volumes of water. For other chemicals, specific disposal instructions, where available, must be followed. These will include collection of specific types of chemical waste in containers for disposal by incineration, landfill, etc.

Storage areas, such as refrigerators, freezers and cupboards, should be regularly checked to avoid build-up of unnecessary items. Reagents and standards, which have passed their expiry date, and samples which need no longer be retained, should be appropriately disposed of. The laboratory should ideally keep records of what it has disposed of and when (and possibly also how).

5.8 Maintenance and Calibration of Equipment

Equipment can reasonably be expected to perform to its full capability when new but may deteriorate rapidly in use unless properly maintained and calibrated. Maintenance of equipment can be either preventive or curative. Some simple maintenance will be possible by the user; however, in many cases it will be the responsibility of the manufacturer, supplier or a recognized agent. Use of this 'professional' maintenance may be a condition of the warranty, and 'do-it-yourself' repairs may invalidate the warranty.

In the case of **preventive maintenance**, the instrument will be the subject of a regular service contract with the frequency of the service depending on the nature of the maintenance. This provides a means of ensuring that the instrument is kept in a general state of 'good health' and identifying any long-term problems. It does not guarantee against sudden breakdown, although sometimes such an arrangement with a manufacturer may provide for preferential service in the event of a breakdown.

Curative maintenance involves calling out engineers only when the instrument has broken down and cannot be repaired by the user. If the user has not exercised routine care when using the instrument, it may have been 'run into the ground' before breaking down. There is thus more responsibility on the user to ensure that the instrument is not abused. Preventive maintenance provides the better means for ensuring instruments are kept in good working order. At first sight, it appears to be the more expensive of the two options. However, in the long term, taking into account factors such as instrument life and time lost during 'downtime', it may prove to be the cheaper option.

Between outside maintenance visits, the laboratory should carry out simple routine maintenance. As a matter of course, the instruments should be kept clean, in particular, spilt chemicals should be cleaned up as soon as possible. Other simple checks that can be carried out within the laboratory will usually be listed in the manufacturer's manual.

Regular calibration and verification ensures that the parameters measured by a particular instrument can be related to a recognized standard. The frequency of instrument calibration may be quite varied, depending largely on the application. If, during the verification of instrument performance, it has been shown that the instrument stays in calibration for about three months, the calibration would be repeated at approximately two-monthly intervals. However, verification (system suitability) will be carried out each time samples are analysed. For some critical analyses, calibration may be performed for each batch of samples or, in an extreme case, for each separate sample.

Modern, microprocessor-controlled instruments often have an 'internal standard', with the instrument undergoing an automatic verification check every time the instrument is used. This may be perfectly satisfactory if the standard can be related to traceable calibration standards. To do this, it is usually necessary to perform a manual confirmation by using an external standard. For example, the internal mass check on an electronic balance can be verified by using a set of calibrated weights. If the confirmation procedure reveals that the instrument is not within its acceptable limits, then some form of corrective action will be required. This may involve an adjustment to the instrument so that it now falls within its correct specification or a maintenance visit may be required.

As a matter of 'good housekeeping', verification procedures should be carefully documented. Where the laboratory is working to a particular quality management standard, there are often strict requirements governing this documentation. Calibration/verification is a very important aspect of making a measurement; the whole process hinges on whether the process is valid. The documentation should include information on the actual procedure and some technical background, indicating when corrective action is necessary and what corrective action should be taken, and how the calibration should be recorded. Calibration/verification records should be carefully and neatly documented, since, as well as providing proof that a system is working, they also indicate when performance is deteriorating and corrective action or maintenance is required. For example, the response of a spectrometer to a particular chemical standard may be fairly constant as long as the instrument is working properly. As a fault develops in the spectrometer's detector, response to the chemical standard changes and this is reflected in the results obtained during the verification process.

A laboratory should draw up a plan of what performance checks are required for each instrument, why they are necessary and how they should be carried out. Maintenance intervals and performance checks for certain instruments are documented in the CITAC/Eurachem Guide to Quality in Analytical Chemistry [14].

Summary

In this chapter, the various aspects of making a measurement have been covered with emphasis on what can cause unacceptable results. It provides guidance as to what constitutes accepted 'best practice' and how this is achieved. Starting from the actions which should be taken before beginning the analysis, it goes on to the actual performance of the analysis and finally what is necessary once the analytical procedures have been completed. Standards, including reference materials, feature strongly in this chapter, as does calibration and verification of instrument performance. Just because a method had been validated does not guarantee that the method will always produce results that are 'fit for purpose'. The reason for this may be down to the person performing the analysis, the equipment or the reagents. The current performance of a method is checked using control samples. Although theoretical aspects of this topic are covered in Chapter 6, the more practical aspects are covered in this present chapter.

This chapter covers briefly the environmental factors which contribute to the reliability of results, including laboratory design, siting of instruments and their maintenance. Mistakes often happen because simple actions are omitted, e.g. containers are not correctly or adequately labelled, incorrect containers are used or instructions are ambiguous.

References

1. 'International Vocabulary of Metrology – Basic and General Concepts and Associated Terms (VIM)', ISO/IEC Guide 99:2007, International Organization for Standardization (ISO), Geneva, Switzerland, 2007.
2. 'Quality Management Systems – Fundamentals and Vocabulary', ISO 9000:2005, International Organization for Standardization (ISO), Geneva, Switzerland, 2005.
3. 'Traceability in Chemical Measurement: A Guide to Achieving Comparable Results in Chemical Measurement', Eurachem, Co-operation on International Traceability in Analytical Chemistry (CITAC), 2003. [http://www.eurachem.org].
4. Barwick, V. and Wood, S. (Eds), *An Analyst's Guide to Meeting the Traceability Requirements of ISO 17025*, 3rd Edition, ISBN 0-948926-23-6, LGC, Teddington, UK, 2005. [http://www.nmschembio.org.uk] (accessed 8 November, 2007).
5. 'Reference Materials – General and Statistical Principles for Certification', ISO Guide 35, International Organization for Standardization (ISO), Geneva, Switzerland, 2006.
6. Emons, H., Fajgelj, A., van der Veen, A. M. H. and Watters, R., *Accred. Qual. Assur.*, **10**, 576–578 (2006).
7. 'General Requirements for the Competence of Reference Materials Producers', ISO Guide 34, International Organization for Standardization (ISO), Geneva, Switzerland, 2000.
8. 'Reference Materials – Contents of Certificates and Labels', ISO Guide 31, International Organization for Standardization (ISO), Geneva, Switzerland, 2000.
9. Bedson, P. and Sargent, M., *Accred. Qual. Assur.*, **1**, 265–274 (1996).
10. Fisher Scientific UK, Loughborough, UK. [http://www.fisher.co.uk] (accessed 4 December, 2006).
11. 'Council Directive 98/24/EC of 7 April, 1998 on the protection of the health and safety of workers from the risks related to chemical agents at work (fourteenth individual Directive within the meaning of Article 16(1) of Directive 89/391/EEC)', *Official Journal of the European Communities*, **41**(L131), 11–23 (1998).
12. 'The Control of Substances Hazardous to Health Regulations 2002', ISBN 0-11-042919-2, Statutory Instrument 2002, No. 2677, Her Majesty's Stationery Office (HMSO), London, UK, 2002.
13. *The Merck Index*, Rahway, NJ, USA. [http://www.merckindex.com] (accessed 4 December, 2006).
14. 'Guide to Quality in Analytical Chemistry, an Aid to Accreditation' (Eurachem), Co-operation on International Traceability in Analytical Chemistry (CITAC), 2002. [http://www.eurachem.org].

Chapter 6

Data Treatment

Learning Objectives

- To understand the key statistical parameters used to describe data sets.
- To understand how to construct and interpret control charts.
- To be aware of the meaning of the terms 'uncertainty', 'error', 'precision', 'bias' and 'accuracy'.
- To appreciate why measurement results are not complete without an estimate of the measurement uncertainty.
- To be able to apply a systematic approach to estimating uncertainty.

This chapter deals with handling the data generated by analytical methods. The first section describes the key statistical parameters used to summarize and describe data sets. These parameters are important, as they are essential for many of the quality assurance activities described in this book. It is impossible to carry out effective method validation, evaluate measurement uncertainty, construct and interpret control charts or evaluate the data from proficiency testing schemes without some knowledge of basic statistics. This chapter also describes the use of control charts in monitoring the performance of measurements over a period of time. Finally, the concept of measurement uncertainty is introduced. The importance of evaluating uncertainty is explained and a systematic approach to evaluating uncertainty is described.

Quality Assurance in Analytical Chemistry E. Prichard and V. Barwick

6.1 Essential Statistics

6.1.1 Populations and Samples

An analyst working for a drinks manufacturing company has to monitor the amount of an artificial sweetener present in batches of a soft drink. Obviously, the analyst cannot test all of the drink that is produced in a particular production batch. As well as being very time-consuming, this approach would not leave any of the product for the manufacturer to sell! The analyst therefore takes a number of *samples* from the batch, analyses these and uses the results to make a judgement about the batch as a whole. In chemical analysis, we are nearly always concerned with making measurements on a relatively small number of samples taken from a much larger number of possible samples. The data we obtain therefore represent a sample from a much larger population of data. We use our sample of data to give us an estimate of the properties of the underlying population of data from which our sample was drawn.

6.1.2 Describing Distributions of Data

An analyst has been asked to determine the cholesterol concentration in a particular tub of low-fat spread. The analyst takes ten samples from the tub and determines the cholesterol concentration in each sample. The results are shown in Table 6.1.

These ten results represent a sample from a much larger population of data as, in theory, the analyst could have made measurements on many more samples taken from the tub of low-fat spread. Owing to the presence of random errors (see Section 6.3.3), there will always be differences between the results from replicate measurements. To get a clearer picture of how the results from replicate measurements are distributed, it is useful to plot the data. Figure 6.1 shows a frequency plot or histogram of the data. The horizontal axis is divided into 'bins', each representing a range of results, while the vertical axis shows the frequency with which results occur in each of the ranges (bins).

There are three things that the analyst may be interested in establishing from the data shown in Figure 6.1, as follows:

Table 6.1 Results from the determination of cholesterol

Cholesterol (mg (100 g)$^{-1}$)	
271.4	268.4
267.8	269.6
268.7	272.5
269.6	270.1
269.7	268.6

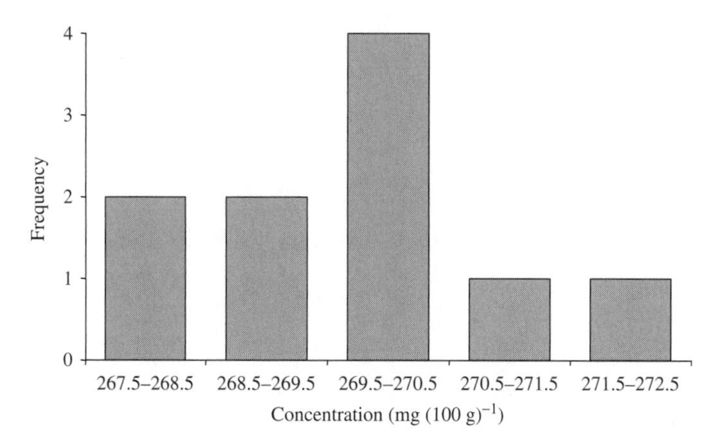

Figure 6.1 Frequency plot (histogram) of the data shown in Table 6.1.

- an estimate of the true value for the parameter being measured (i.e. the actual concentration of cholesterol in the tub of low-fat spread);
- an estimate of the spread of results (i.e. how much do individual results vary from one measurement to the next?);
- an estimate of how much the estimate of the true value may vary from one set of results to another.

Figure 6.2 shows a frequency plot for a much larger sample of data (1000 data points) which was simulated from the data set shown in Table 6.1, using a computer program.

With a larger sample of data, the histogram now gives a much clearer picture of the distribution of the data. Note that the data are concentrated in the central area of the histogram and that the distribution is approximately symmetrical. Finally, with a very large amount of data and a large number of bins, as shown in Figure 6.3, the shape of the underlying population becomes clear. We can now think of the population of data as being described by a smooth curve, the equation of which we could, in principle, determine. The distribution illustrated in Figure 6.3 is known as a 'normal distribution'.

The normal distribution describes the way measurement results are commonly distributed. This type of distribution of data is also known as a *Gaussian distribution*. Most measurement results, when repeated a number of times, will follow a normal distribution. In a normal distribution, most of the results are clustered around a central value with fewer results at a greater distance from the centre. The distribution has an infinite *range*, so values may turn up at great distances from the centre of the distribution although the probability of this occurring is very small.

A normally distributed population of data can be characterized by two parameters. The centre, or location, of the population is described by the parameter μ

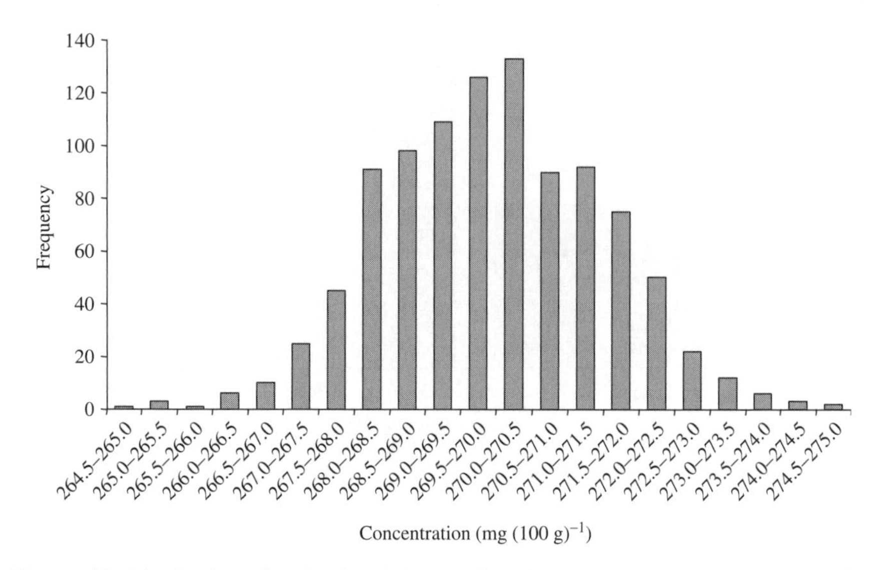

Figure 6.2 Distribution of a simulated data set based upon the data shown in Table 6.1.

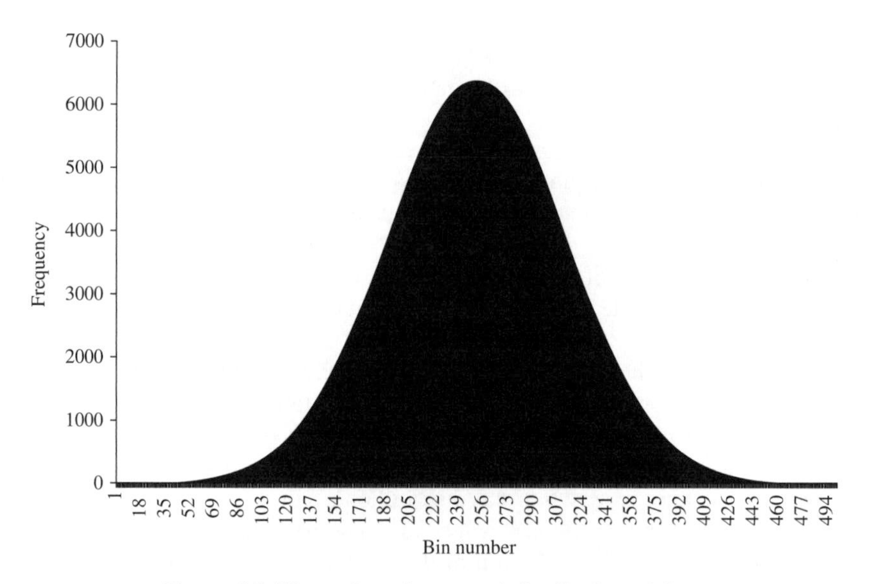

Figure 6.3 Illustration of a normal distribution of data.

(Greek letter, pronounced 'mu'), while its spread is characterized by the parameter σ (Greek letter, pronounced 'sigma'), as shown in Figure 6.4. The parameters μ and σ are known as the *mean* and *standard deviation of the population*, respectively.

In a normal distribution of data, 68.3% of the values lie within $\pm$ 1 standard deviation of the mean value while 95.4% of the values lie within $\pm$ 2 standard

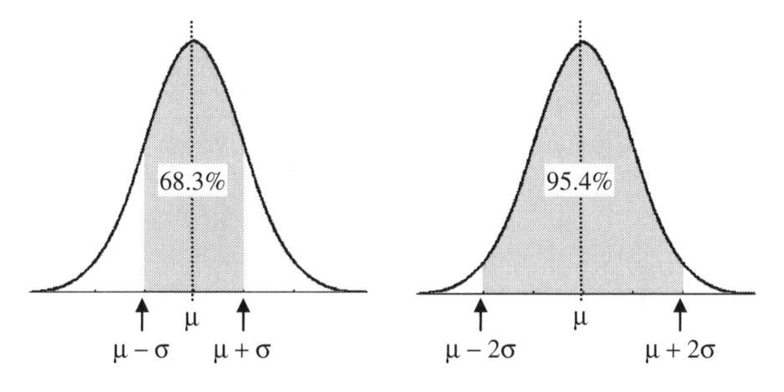

Figure 6.4 Areas under the normal curve.

deviations of the mean value, as shown in Figure 6.4. In addition, 99.7% of the values lie within ± 3 standard deviations of the mean value. Thus, almost all values in a normal distribution are contained within ± 3 standard deviations of the mean value.

6.1.3 Essential Calculations

As mentioned in Section 6.1.1, analysts generally have only a sample of data from a much larger population of data. The sample is used to estimate the properties, such as the mean and standard deviation, of the underlying population.

6.1.3.1 Mean

The *arithmetic mean*, $\overline{x}$, of a sample containing n data points is:

$$\overline{x} = \frac{\sum_{i=1}^{n} x_i}{n} \tag{6.1}$$

If the sample of data is random, then $\overline{x}$ is the best estimate of the population mean, μ.

$$\mu = \frac{\sum_{i=1}^{N} x_i}{N} \tag{6.2}$$

where N is the number of data points in the population.

6.1.3.2 Variance

The *variance of a population*, σ^2, is the mean of the squared deviation of each value from the population mean:

$$\sigma^2 = \frac{\sum_{i=1}^{N} (x_i - \mu)^2}{N} \tag{6.3}$$

The *variance of a sample*, s^2, is:

$$s^2 = \frac{\sum_{i=1}^{n} (x_i - \overline{x})^2}{n - 1} \quad (6.4)$$

6.1.3.3 Standard Deviation

The standard deviation is the square root of the variance.

The *population* standard deviation is given by:

$$\sigma = \sqrt{\frac{\sum_{i=1}^{N} (x_i - \mu)^2}{N}} \quad (6.5)$$

while the *sample* standard deviation is given by:

$$s = \sqrt{\frac{\sum_{i=1}^{n} (x_i - \overline{x})^2}{n - 1}} \quad (6.6)$$

The sample standard deviation, s, provides an estimate of the population standard deviation, σ. The $(n - 1)$ term in equations (6.4) and (6.6) is often described as the *number of degrees of freedom* (frequently represented in statistical tables by the parameter ν (Greek letter, pronounced 'nu'). It is important for judging the reliability of estimates of statistics, such as the standard deviation. In general, the number of degrees of freedom is the number of data points (n) less the number of parameters already estimated from the data. In the case of the sample standard deviation, for example, $\nu = n - 1$ since the mean (which is used in the calculation of s) has already been estimated from the same data.

6.1.3.4 Relative Standard Deviation and Coefficient of Variation

Relative measures of the spread of data are often used, particularly where, for example, the spread of results seems to increase with analyte concentration. The *relative standard deviation* (*RSD*) is a measure of the spread of data in comparison to the mean of the data:

$$RSD = \frac{s}{\overline{x}} \quad (6.7)$$

The relative standard deviation is also known as the *coefficient of variation* (*CV*). The *RSD* is often expressed as a percentage:

$$\%RSD = \%CV = 100 \times \frac{s}{\overline{x}} \quad (6.8)$$

6.1.3.5 Standard Deviation of the Mean

Given the same underlying spread of data (standard deviation, s), as more data are gathered, we become more confident of the mean value, $\overline{x}$, being an accurate representation of the population mean, μ.

As the number of observations, n, in each sample increases, so the standard deviation of the mean values becomes smaller. This is illustrated in Figure 6.5. The *standard deviation of the mean*, $s(\overline{x})$, is the measure of the dispersion of mean values:

$$s(\overline{x}) = \frac{s}{\sqrt{n}} \tag{6.9}$$

where s is the sample standard deviation and n is the number of data points in the sample.

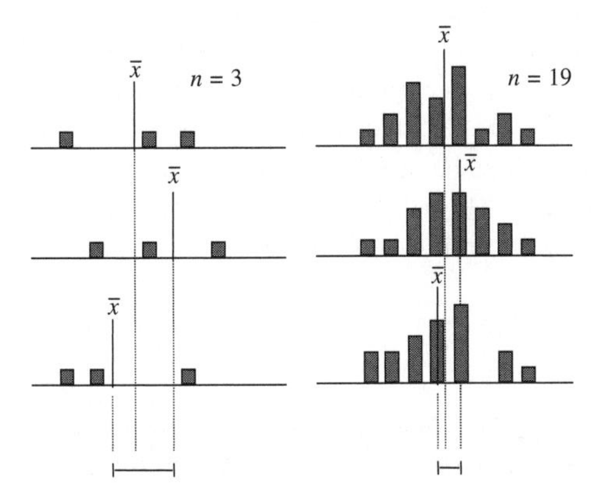

Figure 6.5 Reduction in the dispersion of mean values as the sample number, n, increases.

DQ 6.1

What is the difference between the standard deviation and the standard deviation of the mean?

Answer

The standard deviation is used to describe the dispersion of individual measurement results. If we make a number of repeated measurements on the same sample, the standard deviation provides an estimate of the expected spread of the results. The standard deviation of the mean describes the dispersion of mean values estimated from a number of samples drawn at random from the same population of data. The standard deviation of the mean will always be smaller than the standard deviation by a factor of $\sqrt{n}$, where n is the number of values that have been averaged to obtain the estimate of the mean.

6.1.3.6 Confidence Interval for the Mean

The *confidence interval* gives the range of values within which the true mean is likely to lie, at a stated level of confidence. It is calculated by multiplying the standard deviation of the mean by the appropriate value of $t_{(\nu,\alpha)}$:

$$\bar{x} \pm t_{(\nu,\alpha)} \times \frac{s}{\sqrt{n}} \tag{6.10}$$

where $t_{(\nu,\alpha)}$ is the Student t-value and is obtained from statistical tables (a table of Student t-values is given in the Appendix (see p. 253)). The appropriate value of t depends both on $n - 1$ (i.e. the degrees of freedom) and the level of confidence required. In many statistical tables, the confidence level is given in terms of α, the probability of a value being *outside* the specified range. The confidence level, expressed as a percentage, is equal to $100 \times (1 - \alpha)$. For example, for 95% confidence, α is 0.05. The expression, '**level of significance**' is also used. This is the α-value expressed as a percentage – 95% confidence is therefore equivalent to a significance level of 5%. Knowing the confidence interval, we can say, 'The observed mean might have come from a true mean within this range in a fraction $(1 - \alpha)$ of cases'. Equally, a mean value *outside* the same range would be expected only in a fraction, α, of similar experiments.

SAQ 6.1

For the data set given in Table 6.2, calculate the mean, sample standard deviation, relative standard deviation, degrees of freedom and the 95% confidence interval for the mean.

Table 6.2 Results from the analysis of a Certified Reference Material (BCR CRM 164 anhydrous milk fat certified for cholesterol content)

Cholesterol (mg (100 g)$^{-1}$)	
271.4	269.5
266.3	270.1
267.8	269.7
269.6	268.6
268.7	268.4
272.5	

6.2 Control Charts

As mentioned in Chapter 5, Section 5.4.2, the analysis of quality control samples is an important aspect of quality control. The analysis of quality control samples allows the analyst to monitor the performance of a measurement system over a period of time. This can generate a great deal of data, but the monitoring is only worth doing if the data so produced can be interpreted. One of the most useful ways of looking at the data is to plot a *control chart*. The user can define warning and action limits on the chart to act as 'alarm bells' for when the system is going out of control. A control chart is simply a chart on which measured values of whatever is being measured are plotted in time sequence, for instance, the successive values obtained from measurement of the quality control sample. By plotting this information on a chart, a graph is produced in which the natural fluctuations of the measured value can readily be appreciated.

This book has only limited scope for describing control charts and the statistical theory on which they are based. Some simple applications are briefly described below, together with a simplified statistical explanation. For more detailed information, you should refer to the relevant standards and guides [1–5].

As has already been mentioned, successive measurements of a characteristic by using a particular method will show a natural variation arising from random errors associated with the method. The set of results will have a *mean* value and most commonly, the values will be symmetrically distributed around the mean in a *normal distribution* (see Section 6.1.2). As shown in Figure 6.4, approximately 95% of values in a normal distribution will lie within $\pm$ 2 standard deviations of the mean value and 99.7% will lie within $\pm$ 3 standard deviations of the mean value. It is therefore unlikely (5% probability) that a member of a normally distributed data set will be further away from the mean than 2 standard deviations, and very unlikely (0.3% probability) that it will be further away from the mean than 3 standard deviations. Further measurements should behave in the same way and lie within these boundaries. If they do not, then it is probable that some change has occurred to the measurement system which has significantly altered its performance, thus causing a shift in the mean or an increase in the standard deviation. The purpose of the control chart is to make this change evident. The user must decide whether or not this change is significant.

6.2.1 The Shewhart Chart

This is the simplest type of control chart. It is typically used to monitor day-to-day variation of an analytical process. It does so by monitoring the variation of a quality control (QC) sample or standard when measured by the process. Measurement value is plotted on the y-axis against time or successive measurement on the x-axis. The measurement value on the y-axis may be expressed as an absolute value or as the difference from the target value. The QC sample is a sample typical of the samples usually measured by the analytical process, which is stable and available in large quantities. This QC sample is analysed at appropriate regular intervals in the sample batches. As long as the variation in the measured result for the QC sample is acceptable, it is reasonable to assume that the measured results for test samples in those batches are also acceptable. However, how do we determine what is acceptable and what is not?

We do this by using the statistical ideas outlined above. First of all, the QC sample is measured a number of times (under a variety of conditions which represent normal day-to-day variation). The data produced are used to calculate an average or mean value for the QC sample, and the associated standard deviation. The mean value is frequently used as a '**target**' value on the Shewhart chart, i.e. the value to 'aim for'. The standard deviation is used to set action and warning limits on the chart.

Once the chart is set up, day-to-day QC sample results are plotted on the chart and monitored to detect unwanted patterns, such as 'drift' or results lying outside the warning or action limits. In Figure 6.6, Shewhart charts have been used to show four types of data: (a) data subject to normal variation, (b) as in (a) but offset from the target value, (c) gradual drift and (d) step-change. To keep things simple, action and warning limits have only been included in (a).

It is normal to use warning limits at $\pm$ 2 standard deviations and action limits at $\pm$ 3 standard deviations from the target value. From the statistical rules already described, we would expect very few results (i.e. 3 in 1000) to fall outside the action limits and 1 in 20 to fall outside the warning limits. These limits apply if each value plotted on the chart represents a single measurement of the QC sample. If the method being monitored requires the analysis of test samples in replicate and a mean value to be reported, then the QC sample should be treated in the same way. For example, if the analytical process being monitored requires test samples to be analysed in duplicate and the mean of the two results to be reported, the QC sample should be analysed in duplicate and the average value plotted on the control chart. Plotting mean values rather than single results removes some of the variation from the data and so we need to make changes to the action and warning limits if they are to have any practical use for monitoring the data. When mean values are plotted on a Shewhart chart, the warning and action limits should be based on the standard deviation of the mean, rather than the standard deviation (see Section 6.1.3). The action and warning limits should therefore be plotted at $\pm\ 3\sigma/\sqrt{n}$ and $\pm\ 2\sigma/\sqrt{n}$ units of standard deviation, respectively.

When using control charts, you should take action on any points which fall outside the action limits and be alert when points exceed the warning limits. There are three other situations which normally indicate a problem with the system, as follows:

(i) three successive points outside the warning limits but inside the action limits;

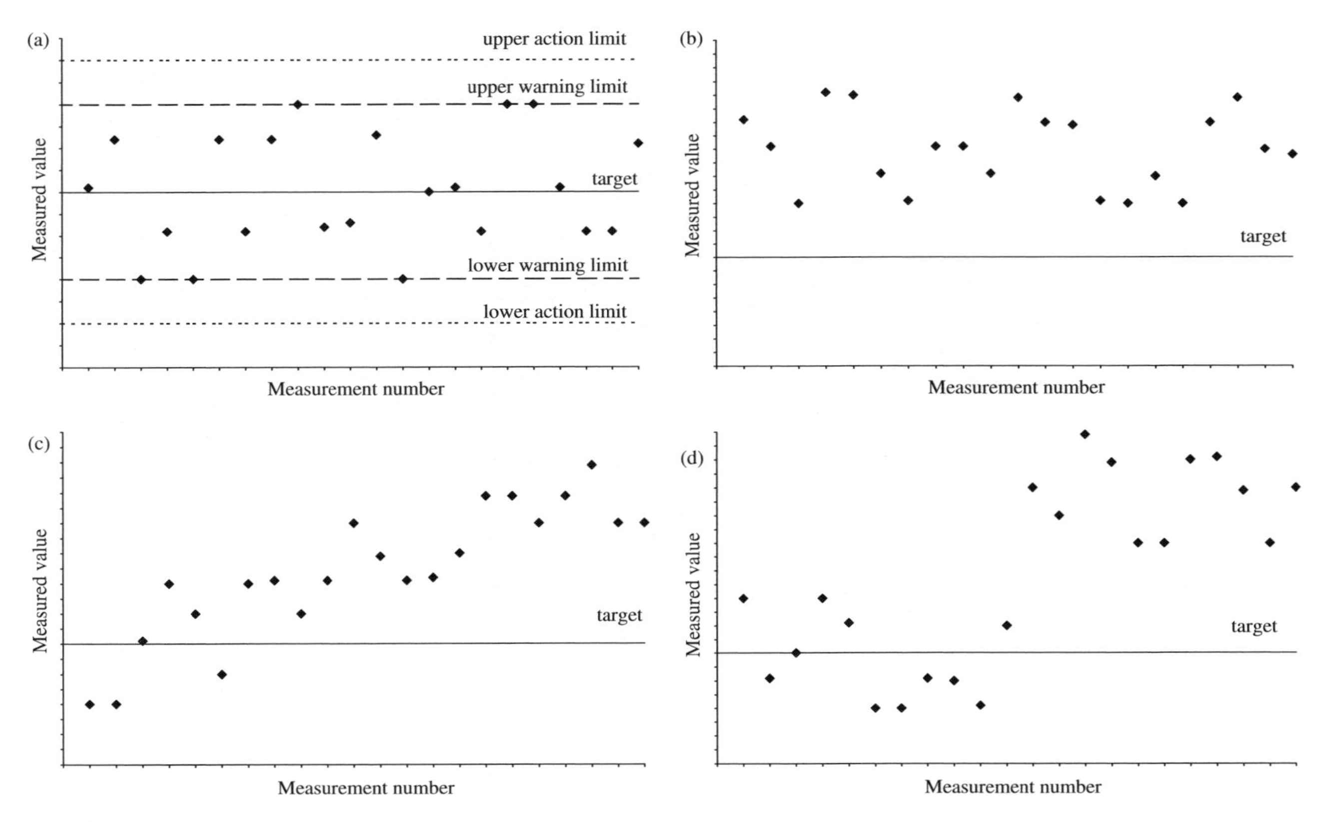

Figure 6.6 Shewhart charts showing: (a) data 'in control' about the target value; (b) data offset from the target value; (c) drifting data; (d) data with a step-change.

(ii) two successive points outside of the warning limits but inside the action limits on the same side of the mean;

(iii) ten successive points on the same side of the mean.

These are the basic Shewhart rules, ISO 8258:1991 gives additional rules for identifying abnormal patterns in the data [1].

6.2.2 Moving Average Chart

One disadvantage of the Shewhart chart is that progressive changes or step-changes do not readily stand out from the natural variation inherent in the method. A slightly different chart, called the ***moving average chart***, alleviates this problem by averaging out natural variation before plotting so that only the significant changes are evident. It works by averaging values (typically four) in succession.

For example, in a moving average chart where values are averaged four at a time ($n = 4$), measurements 1, 2, 3 and 4 are averaged and plotted as the 1st point, 2, 3, 4 and 5 are averaged and plotted as the 2nd point, 3, 4, 5 and 6 are averaged and plotted as the 3rd point, etc.

In Figure 6.7, the same data, containing an undesirable step-change, have been plotted on (a) a Shewhart chart and (b) a moving average chart ($n = 4$). Compare the two and note how in (b) the step-change is much more obvious against the background variation. In Figure 6.7(a), the step-change in results does not become clear until measurements 12 and 13. However, in Figure 6.7(b) the moving average starts to increase after measurement number 10.

The user can vary 'n' to suit. The larger the value of 'n', the better the smoothing effect on the data but the longer the response time before significant changes are evident. So, for a particular application the user has to balance the response time for the highlighting of change against the degree of smoothing required.

If action and warning limits are added to a moving average chart, they should be plotted at $\pm\, 3/\sqrt{n}$ and $\pm\, 2/\sqrt{n}$ units of standard deviation, respectively.

6.2.3 CUSUM Charts

Because it uses all of the data, the CUSUM chart is the best way of detecting small changes in the mean. Consider a process for which there is a known target value, T. For each new measurement, the difference between the measurement and T is calculated and added to a running total. This running total is plotted against successive measurements (CUSUM is short for cumulative sum).

If the measurement system is operating such that the operating mean is close to the established mean or target value, the gradient of the CUSUM will be close to zero. A positive gradient implies an operating mean greater than the target value, while a negative gradient implies an operating mean less than the target

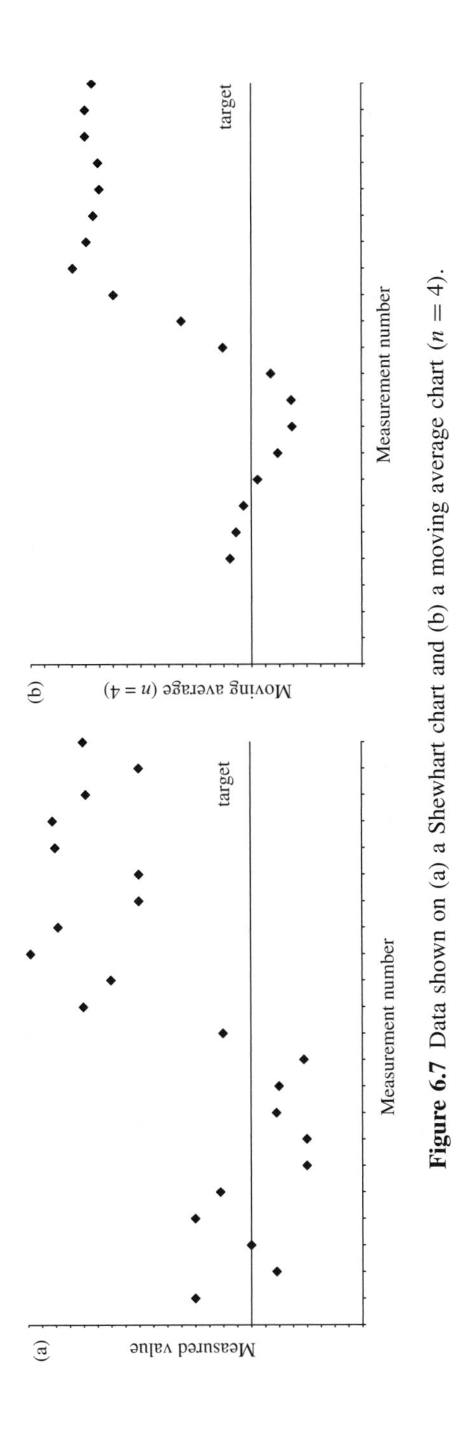

Figure 6.7 Data shown on (a) a Shewhart chart and (b) a moving average chart ($n = 4$).

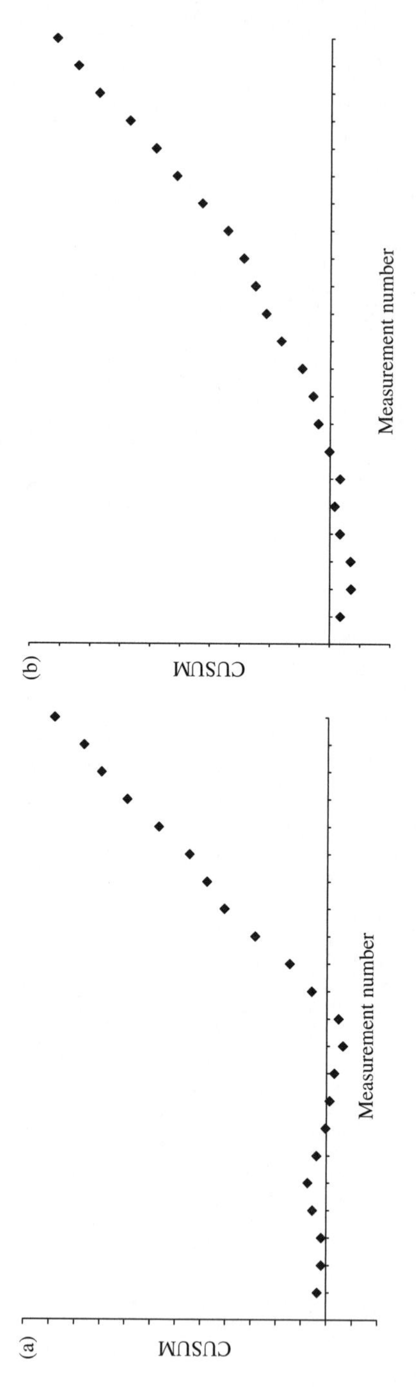

Figure 6.8 CUSUM charts showing (a) a step-change and (b) drift.

value. A step-change in a set of data shows up in a CUSUM as a sudden change of gradient (see Figure 6.8(a)). Note that this CUSUM chart was constructed from the data plotted in Figure 6.6(d). Gradual drift in a system causes small but continuous changes to the mean. In a CUSUM chart, this translates into a constantly changing gradient, i.e. a curve (see Figure 6.8(b)). This chart was constructed from the data plotted in Figure 6.6(c).

Conventional warning and action limits are unsuitable for interpreting whether or not CUSUM data are 'under control'. Instead, we use something called a 'V-mask'. This is usually made of transparent material so that it can be overlaid on the CUSUM chart. A diagram of a 'V-mask' is shown in Figure 6.9(a). The data on the CUSUM chart are examined by laying the mask over the data with the left-hand end of the line 'd' aligned with each data point in turn. The line 'd' is always kept parallel to the x-axis. So long as all of the preceding data lie within the arms of the mask (or their projections), the system is 'in control'. When preceding data points fall outside of the arms of the mask, the system has fallen 'out of control'. Figure 6.9(b) illustrates the use of a 'V-mask' on CUSUM data which are actually subject to drift.

At point A in Figure 6.9(b), all of the preceding data clearly lie within the arms of the mask – the data are therefore 'under control'. However, at point B some of the preceding data points lie below the lower arm of the mask, so indicating that the system is 'out of control' at that point.

The limits of control for a CUSUM chart are defined by the length of d, the lengths of the 'arms' and the angle θ, and consequently these must be chosen with care. The aim is to be able to identify quickly when the system has gone 'out of control' (i.e. when measurement results deviate significantly from the target value) but to avoid too many 'false alarms'. The scales used on the x- and y-axes also have a significant influence on the choice of d and θ. The axes should be scaled so that the divisions on both axes are the same length. A division on the x-axis should represent a single unit, while a division on the y-axis should be equivalent to 2σ (σ is the population standard deviation or an estimate

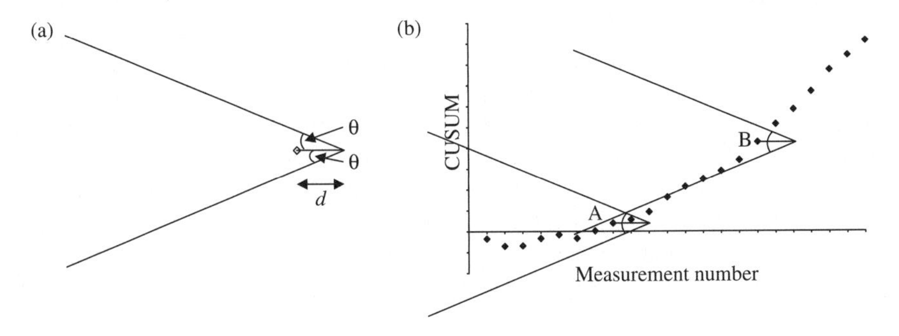

Figure 6.9 (a) The V-mask – for interpretation of CUSUM charts and (b) a CUSUM chart illustrating the use of a V-mask.

obtained from the sample standard deviation), if the values used are from single measurements. If n repeat measurements are taken each time and averaged, then this becomes $2\sigma/\sqrt{n}$. The dimensions of the 'V-mask' can be chosen by trial and error, by examining historical data. The mask should be designed so that you get an 'early warning' that the data have gone 'out of control'. In practice, it should be possible to construct a mask that gives you the same statistical probability of control as the conventional warning and action limits used on a Shewhart chart.

If historical data are not available, then a starting point is to construct a mask with $d = 2$ units of length of the horizontal axis and θ at $\approx 22^\circ$ (assuming the axes have been scaled as described above). The arms are such that the total length, in the horizontal direction, is 12 units. Points within the mask indicate that the system is 'under control' while points outside of the mask indicate that the system is 'out of control'.

6.2.4 Range Charts

The charts described in the previous sections are primarily used for detecting bias in a measurement system. Bias is indicated by shifts in the results from the QC sample. Examples are shown in Figure 6.6(b) (results offset from the target value), Figure 6.6(c) (drifting data) and Figure 6.6(d) (a step-change in the data). However, it is also useful to monitor the precision of the measurement process. The precision is a measure of the variation in results from one measurement to the next. A method may produce results which are unbiased, but the precision of the results may not be acceptable (i.e. the results from repeated measurements show significant variation). A *range chart* can be used to monitor the precision of results. Each time the QC sample is analysed, it must be analysed in replicate. For example, the QC sample could be analysed three times within each batch of test samples. It is important that the replicates are completely independent, i.e. the QC sample must be put through the whole analytical method, not just replication of the end measurement. For each set of replicates, the range (i.e. the difference between the highest and the lowest value) is calculated. The ranges are then plotted sequentially on a chart, as shown in Figure 6.10. As for the Shewhart chart, control limits can be added to a range chart. First, the mean of the ranges is plotted on the chart. Upper and lower action limits are calculated by multiplying the mean range by the appropriate value from Table 6.3. Additional values can be found in ISO 8258:1991 [1]. The multiplier used depends on the number of replicates used to calculate each range value.

Figure 6.10 shows a range chart with upper and lower control limits. In this example, each batch of test samples contained four replicates of the QC sample. The range was calculated for each set of four QC results and plotted sequentially on the chart. The mean range value was 2.7. Since $n = 4$, the lower control limit is set at zero while the upper limit is set at $2.282 \times 2.7 = 6.2$. Note that each batch of analyses must contain the same number of replicates of the QC sample.

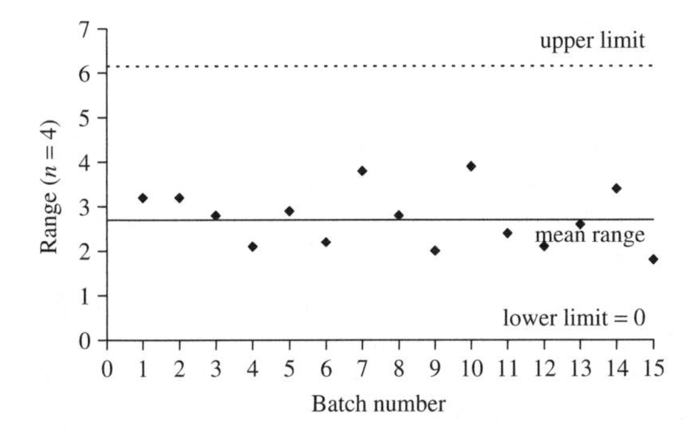

Figure 6.10 Range chart showing upper and lower limits.

Table 6.3 Multipliers for calculating control limits for a range chart with a lower limit = 0

Number of replicates (n)	Upper limit
2	3.267
3	2.574
4	2.282
5	2.114

DQ 6.2

What is the difference between a Shewhart chart and a range chart?

Answer

A Shewhart chart is used to monitor the variation of individual results over time, compared to a target value. Shewhart charts are useful for identifying when bias has entered a measurement system. Non-random patterns in the data, such as drift or step-changes, indicate that bias is present. A range chart is used to monitor the precision of a measurement system, regardless of whether there is any bias present. The data on both types of chart are best evaluated by setting control limits.

SAQ 6.2

Use the following data to construct:

(a) a Shewhart chart;

(b) a moving average chart ($n = 5$);

(*continued overleaf*)

SAQ 6.2 *(continued)*

(c) a CUSUM chart.
The average/target value has already been established from previous data and is equal to 17 and the standard deviation is 1.5.
Data: 16, 16, 18, 14, 16, 15, 18, 17, 18, 18, 16, 18, 15, 16, 17, 21, 17, 21, 20, 22, 19, 19, 21, 22, 20, 21, 20, 19, 22, 21, 21, 21, 22, 21, 21.

(d) On the Shewhart chart, produced in (a), add warning limits at ± 2 standard deviations about the average and action limits at ± 3 standard deviations about the average. At which datum point should the analyst intervene because the system is going 'out of control'?

(e) On the moving average chart, add warning limits at ± $(2/\sqrt{n})$ standard deviations about the average and action limits at ± $(3/\sqrt{n})$ standard deviations about the average. At which datum point should the analyst intervene because the system is going 'out of control'?

(f) On the CUSUM chart, at which data point does a significant and prolonged change in the gradient become evident? Construct a 'V-mask' which will show a loss of control from point 16 onwards.

6.3 Measurement Uncertainty

6.3.1 The Measurement Process

In general, whenever any quantitative measurement is made the value obtained is only an estimate of the *true* value of the property being measured. Many factors will cause the value obtained to differ from the true value. These can be summarized as follows:

(a) imperfections in the measuring device;

(b) imperfections in the measurement method;

(c) operator effects.

The result of a quantitative chemical measurement is not an end in itself. It has a cost and therefore it always has a purpose. It may be used, for example, in checking products against specifications or legal limits, to determine the yield of a reaction, or to estimate monetary value.

Whatever the reason for obtaining it, the result of a chemical measurement has a certain importance since decisions based upon it will very often need to be made. These decisions may well have implications for the health or livelihood of millions of people. In addition, with the increasing liberalization of world trade, there is pressure to eliminate the replication of effort in testing products moving across national frontiers. This means that quantitative analytical results should be acceptable to all potential users whether they are inside or outside of the organization or country generating them.

It is clear then that some indicator of quality is required if chemical measurements are to be used with confidence. Such an indicator must:

(a) be universally applicable;

(b) be consistent;

(c) be quantifiable;

(d) have a meaning that is clear and unambiguous.

An indicator that meets these requirements is *measurement uncertainty*.

6.3.2 Definition of Uncertainty

Uncertainty is a parameter that characterizes the range of values within which the value of the quantity being measured is expected to lie. This means that the result of a quantitative measurement cannot properly be reported as a single value, e.g. pH = 3.7, as there will always be some doubt about the measured value. We cannot be certain that the single value obtained at the end of a measurement process is the *true value*. We can, however, have more confidence in our result if we regard the value obtained as an estimate, since this is a weaker assertion. Of course, simply lowering the status of our result in this way would not be very helpful to anyone wanting to use it. A potential user of our result would really be interested in the true value of the quantity being measured. However, true value is a hypothetical concept since it cannot be measured. The best we can do is to report a range of values, centred on our estimate of the true value, and state that the true value lies somewhere within this range. Calculating this range is what measurement uncertainty is all about.

6.3.3 Errors

Uncertainty and *error* are two quite distinct concepts and should not be confused.

An **error** is the difference between an individual result and the true value of the quantity being measured. Since true values cannot be known exactly, it follows, from the above definition, that errors cannot be known exactly either.

Errors are usually classified as either *random* or *systematic*.

DQ 6.3

Write down what you think the difference is between random and systematic errors.

Answer

Random error is the result of chance variation and causes results to vary in an unpredictable way. *Systematic error* causes results to be consistently higher (or lower) than expected. These concepts are discussed in more detail in the following sections.

6.3.3.1 Random Error

Random error arises as the result of chance variations in factors that influence the value of the quantity being measured but which are themselves outside of the control of the person making the measurement. Such things as electrical noise and thermal effects contribute towards this type of error. Random error causes results to vary in an unpredictable way from one measurement to the next. It is therefore not possible to correct individual results for random error. However, since random error should sum to zero over many measurements, such an error can be reduced by making repeated measurements and calculating the mean of the results.

6.3.3.2 Systematic Error

In contrast, a systematic error remains constant or varies in a predictable way over a series of measurements. This type of error differs from random error in that it cannot be reduced by making multiple measurements. Systematic error can be corrected for if it is detected, but the correction would not be exact since there would inevitably be some uncertainty about the exact value of the systematic error. As an example, in analytical chemistry we very often run a 'blank' determination to assess the contribution of the reagents to the measured response, in the known absence of the analyte. The value of this blank measurement is subtracted from the values of the sample and standard measurements before the final result is calculated. If we did not subtract the blank reading (assuming it to be non-zero) from our measurements, then this would introduce a systematic error into our final result.

Where the value of a systematic error is known, or can be calculated, it should be corrected for. Any correction we make is unlikely to be exact and so we must also produce an estimate of the amount by which our correction could be wrong. This estimate must be included in our uncertainty calculations.

Figure 6.11 illustrates the difference between these two main types of error, using the example of delivering liquid from a 25 ml Class A pipette.

The pipette has a stated volume of 25 ml. However, due to the manufacturing process the actual volume of liquid from a particular pipette filled to the calibration mark (ignoring any random errors) is found to be 25.02 ml. This is within the permitted tolerance for a 25 ml Class A pipette ($\pm$ 0.03 ml according to BS 1583:1986 [6]). This is a systematic error, as the volume of liquid delivered from the pipette will always be 0.02 ml greater than the stated volume each time the

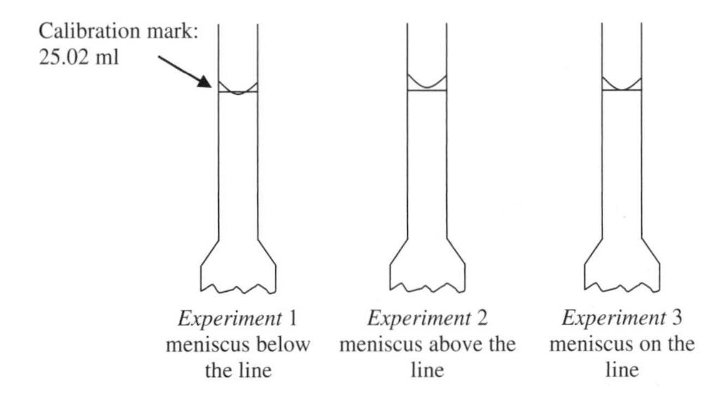

Figure 6.11 Illustration of the differences between random and systematic errors, using the example of delivering liquid from a 25 ml Class A pipette.

pipette is used. In addition, every time an analyst uses the pipette the exact position of the meniscus in relation to the calibration line will vary slightly. This is a random error, as the position of the meniscus will vary in an unpredictable way. This is illustrated in Figure 6.11 by the small change in the position of the meniscus from one experiment to the next. The presence of both of these types of error means that the volume of liquid delivered from the pipette will vary slightly each time the pipette is used. This leads to the uncertainty in the volume of liquid delivered by the pipette.

For any given measurement process, more than one instance of each type of error can apply. Therefore, errors are insufficient to describe the quality of a measurement result. Measurement uncertainty, on the other hand, combines into a single range the effect of all of the different factors that can influence a measurement result.

If you identified most of the differences between random and systematic errors correctly, you obviously have a good understanding of the nature of error in chemical measurement. If you had difficulty with this do not worry, but now is the time to get these ideas clear in your mind.

6.3.4 Precision, Bias and Accuracy

The concepts of precision, bias and accuracy were introduced in Chapter 4. However, as they are important in the context of evaluating measurement uncertainty it is worth revisiting them.

Precision is the closeness of agreement between independent test results obtained under stipulated conditions. The precision tells us by how much we can expect the results of repeated measurements to vary. The precision of a set of measurement results will depend on the magnitude of the random errors affecting the measurement process. Precision is normally expressed as a standard deviation or relative standard deviation (see Section 6.1.3).

Bias is the difference between the mean value of a number of test results and an accepted reference value. The magnitude of the bias will depend on the size and direction of systematic errors.

Bias is a measure of 'trueness'. It tells us how close the mean of a set of measurement results is to an assumed true value. Precision, on the other hand, is a measure of the spread or dispersion of a set of results. Precision applies to a set of replicate measurements and tells us how the individual members of that set are distributed about the calculated mean value, **regardless of where this mean value lies with respect to the true value.**

Accuracy is the closeness of the agreement between the result of a measurement and the true value of the quantity being measured. Accuracy is the property of a *single* measurement result. It tells us how close a single measurement result is to the true value and therefore includes the effect of both precision and bias.

Figure 6.12 illustrates the difference between precision, bias and accuracy. In examples (a) and (b), there is no bias. However, the precision in case (b) is better than in case (a) (i.e. the dispersion of results is smaller). Individual results obtained in case (b) would therefore be considered to be more accurate than those

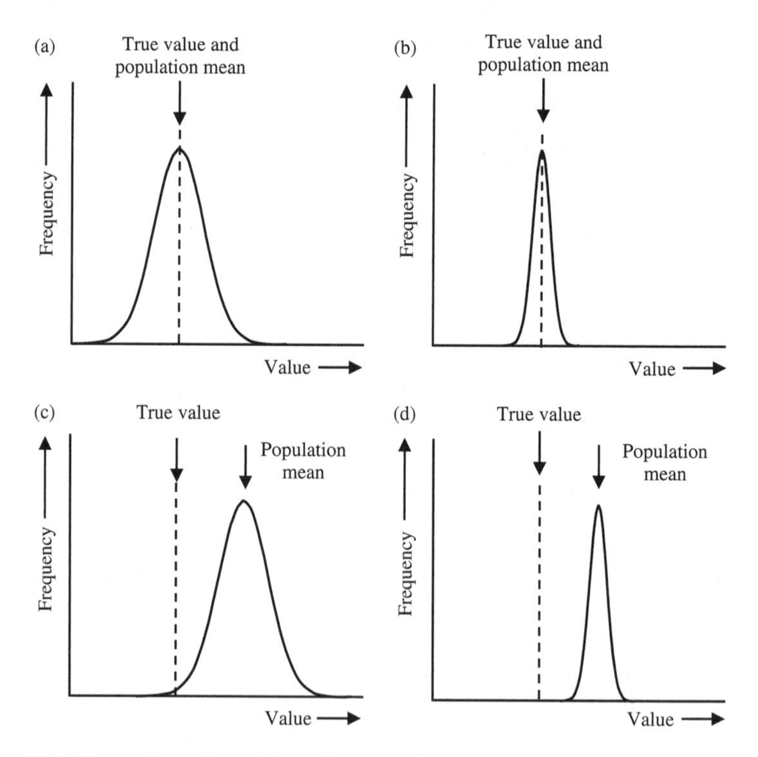

Figure 6.12 Illustration of precision, bias and accuracy: (a) not biased, not precise; (b) not biased, precise, accurate; (c) biased, not precise; (d) biased, precise.

obtained in case (a). In cases (c) and (d), the results are biased. The precision in case (d) is the same as in case (b) but the results in case (d) are inaccurate as they differ significantly from the true value.

Two other terms that you will come across when working with chemical data are *repeatability* and *reproducibility*. Again, these two terms can easily be confused and you should learn to distinguish between them. They are both measures of precision.

DQ 6.4

Write down your definitions of repeatability and reproducibility.

Answer

Repeatability refers to the variability in a series of results obtained for a given measurement carried out:

- by the same operator;
- using the same equipment;
- in the same laboratory;
- at a particular time.

Reproducibility refers to the variability in a series of results obtained for a given measurement carried out:

- by different operators;
- using different equipment;
- in different laboratories;
- at different times.

Repeatability and reproducibility are discussed in more detail in Chapter 4.

SAQ 6.3

Which of the following statements correctly describes uncertainty?

(a) A type of error.

(b) A measure of precision.

(c) The reciprocal of accuracy.

(*continued overleaf*)

SAQ 6.3 *(continued)*

(d) A range of values containing the true value.

(e) The range of values between the true value and a measured value.

(f) The odds against getting the correct result.

6.3.5 Evaluating Uncertainty

As we have seen in previous sections, the result of a measurement is not complete unless an estimate of the uncertainty associated with the result is available. In any measurement procedure, there will be a number of aspects of the procedure that will contribute to the uncertainty. Uncertainty arises due to the presence of both random and systematic errors. To obtain an estimate of the uncertainty in a result, we need to identify the possible sources of uncertainty, obtain an estimate of their magnitude and combine them to obtain a single value which encompasses the effect of all the significant sources of error. This section introduces a systematic approach to evaluating uncertainty.

The uncertainty evaluation process can be broken down into four stages: ***specification***, ***identification***, ***quantification*** *and* ***combination***.

6.3.5.1 Specification

To allow the uncertainty to be evaluated effectively, a 'model equation' describing the method of analysis is required. The starting point is the equation used to calculate the final result. Intially, we will need to consider the uncertainties associated with the parameters that appear in this equation. It may be necessary to add terms to this equation (i.e. expand the model) to include other parameters that may influence the final result and therefore contribute to the measurement uncertainty.

It is also essential to have a clear understanding of the analyte or property being measured. For example, an analyst may be studying the amount of lead present in paint used on toys. One possibility would be to use a method which determines the total amount of lead present. Alternatively, the analyst may be interested in the amount of lead that is released from a paint sample taken from a toy when it has been extracted with a stomach-acid simulant. In both cases, the end measurement is the same – the concentration of lead in a solution. However, the results from the two approaches would be very different. In the first case, the sample will have been digested with a strong acid solution which should release all of the lead present in the sample. In the second case, we would expect the results to be lower as the method is designed to estimate the amount of lead released under particular conditions. The second type of method is sometimes referred to as an *empirical method*. This is a method where the result produced is entirely dependent on the analytical method. In the above example, if the concentration of the stomach-acid simulant or the extraction conditions were changed, the result would change. Other examples of empirical methods include the determination of moisture or fibre in foodstuffs.

When evaluating uncertainty, it is important to understand the distinction between empirical and non-empirical methods, as this influences how the uncertainty is evaluated. In the case of non-empirical methods, any bias in the results which is due to the method of analysis or, for example, a particular sample type, needs to be considered as part of the uncertainty evaluation process. For example, if a method was intended to determine the total amount of cadmium present in a soil sample, but for some reason only 90% of the cadmium present was extracted from the sample, then this 10% bias would need to be accounted for in the uncertainty estimate. One approach would be to correct results to take account of the bias. However, there would be an uncertainty associated with the correction as there will be some uncertainty about the estimate of the bias. For empirical methods, the method bias is, by definition, equal to zero (the method defines the result obtained). However, when evaluating the uncertainty associated with results obtained from an empirical method, we still need to consider the uncertainty associated with any bias introduced by the laboratory during its application of the method. One approach is to analyse a reference material that has been characterized by using the same empirical method. If no suitable reference material is available, then any bias associated with carrying out the individual stages of the method in a particular laboratory will need to be evaluated.

6.3.5.2 Identification

The next stage of the process is to develop a comprehensive list which identifies all of the possible sources of uncertainty in the measurement procedure. At this stage, it is not necessary to worry about the magnitudes of the different sources of uncertainty. Some of the sources of uncertainty identified at this stage may turn out to be insignificant, but this will be addressed in the next stage (quantitation). To help identify the sources of uncertainty for a particular method, you may find it helpful to produce a flow diagram, which sets out the individual steps in the method. You can then identify the sources of uncertainty associated with each step. Remember that you must consider the uncertainty associated with each of the parameters that appear in the model produced at the specification stage, but there will almost certainly be other sources of uncertainty.

DQ 6.5

Make a list of as many of the potential sources of uncertainty as you can think of that could affect the result of a chemical measurement.

Answer

There are many potential sources of uncertainty. Examples include: sampling; the nature of the sample matrix; sample storage conditions;

instrument/equipment effects; reagent purity; measurement conditions; sample preparation; calibration model; operator effects; random effects. These common sources of uncertainty are considered in more detail below.

Sampling If sampling forms part of the specification produced in the first stage of the evaluation process (i.e. the sampling and analysis are performed by the same laboratory), then uncertainties associated with variations between different samples or any bias in the sampling procedure need to be considered. If the laboratory is analysing the sample on an 'as-received' basis, then the effect of sampling carried out away from the laboratory does not have to be considered. However, the uncertainties associated with any sampling that occurs within the laboratory in order to obtain the test portion will need to be considered.

Sample Effects The recovery of an analyte from a complex matrix may be affected by other components of the matrix. The homogeneity of the sample will also influence the results. This is related to the issue of sampling mentioned above. Physical or chemical form can lead to incomplete recovery of the analyte. For example, an element may exist in more than one oxidation state in a sample and hence be incompletely determined by a method that requires it to be in one particular state only (*speciation*). The sample and/or analyte may be unstable, causing a change in the composition of the sample during the course of the analysis.

Storage Conditions If samples are stored for a period prior to analysis, the storage conditions may influence the results. The storage conditions and storage time should therefore be listed as possible sources of uncertainty.

Instrument/Equipment Effects Examples include the calibration and precision of an analytical balance, the specified tolerance for volumetric glassware and a temperature controller that maintains a mean temperature which is different (within specification) from its indicated value.

Reagent Purity The purity of many reagents is specified by the manufacturer as being *not less than* a specified value. Any assumptions about the degree of purity will contribute to the uncertainty. The nature of any impurities may also be important. Pure substance Certified Reference Materials will have a stated purity, plus an estimate of the uncertainty associated with the purity value.

Measurement Conditions If the environmental conditions (temperature and humidity) have a significant effect on the measurement result, then this needs to be addressed by introducing tighter controls. Using items of volumetric glassware at a temperature different from that at which they were calibrated will be a source of uncertainty. This should be evaluated when estimating the uncertainty associated with volumes measured using volumetric glassware.

Sample Preparation Effects Many methods require the sample to be treated in some way before the analyte can be determined. Examples include drying, grinding or blending of the sample, and extraction or digestion of the sample. Variations in the conditions under which these activities are carried out (e.g. extraction temperature and time, solvent composition) may affect the final result.

Operator Effects An analyst may read a meter or scale consistently high or low. There will be some variability in judging the end-point of a titration.

Computation Effects Selection of the calibration model will influence results. For example, fitting a linear calibration function to data that are non-linear will result in increased uncertainty in values predicted by using the calibration function.

Random Effects Random effects (see Section 6.3.3) will contribute to the uncertainty in all measurement procedures. Random effects should therefore always appear in the list of sources of uncertainty.

Your list should have included at least some of the items mentioned above, but you may well have identified other sources of uncertainty. Remember that uncertainty is not about *mistakes*. The uncertainty estimate is intended to reflect the likely variation in results when a method is carried out correctly and operating under statistical control. Your list of sources of uncertainty should not therefore include any gross errors such as contamination of samples, mistakes in calculations or the analyst failing to follow the standard operating procedure correctly.

6.3.5.3 Quantification

The uncertainties identified in the previous stage must now be quantified. There are two recognized approaches to achieving this, as follows:

(i) Evaluate the uncertainty arising from each of the individual sources of uncertainty and combine them by using the mathematical rules described later (this is sometimes referred to as the 'bottom-up' approach to uncertainty estimation).

(ii) Use method performance data, such as precision and bias estimates (measured with respect to an appropriate reference value), to evaluate the combined effect of a number of the identified sources of uncertainty (this is sometimes referred to as the 'top-down' approach). Such estimates can be obtained from method validation studies and quality control data.

In practice, a combination of these approaches may be used.

All uncertainty estimates must be expressed numerically as a ***standard uncertainty***. The term 'standard uncertainty' is analogous to the statistical term ***standard deviation***. A standard uncertainty is therefore an uncertainty estimate expressed as a standard deviation. Standard uncertainties are usually denoted by the symbol u.

Uncertainty estimates can be divided into two categories [7]:

- *Type A*: obtained from the statistical analysis of results from repeated measurements.
- *Type B*: obtained by means other than statistical analysis of results (e.g. data from calibration certificates, manufacturers' specifications, modelling, etc.).

Note that the categories relate only to how the estimate was obtained, and not to whether the uncertainty is due to a random or a systematic effect. Type A uncertainty estimates are, by definition, expressed as a standard deviation. Type B uncertainty estimates can take a number of different forms, and may need to be converted to a standard uncertainty prior to combination with other uncertainty estimates. This is discussed later in this section.

The standard uncertainty arising from random effects is typically measured from precision studies and is quantified in terms of the standard deviation of a set of measured values. For example, consider a set of replicate weighings performed in order to determine the random error associated with a weighing. If the true mass of the object being weighed is 10 g exactly, then the values obtained might be as follows:

$$10.0001, 10.0000, 10.0002, 10.0002, 10.0001, 10.0000, 10.0001,$$
$$10.0000, 10.0002, 10.0000$$

There are 10 values in this set.
The mean value is 10.000 09 g
The standard deviation is, $s = 0.000\ 087\ 559$ g

The calculated standard deviation is now regarded as a standard uncertainty and is expressed as:

$$u(w) = 0.000\ 087\ 559\,\text{g}$$

where $u(w)$ represents the standard uncertainty in the result (w) of a single weighing, arising from random errors associated with the weighing process. If this uncertainty estimate was to be used in further uncertainty calculations, then all of the significant figures should be retained. However, when reporting an uncertainty estimate the value is normally rounded to a maximum of two significant figures. In this example, the uncertainty would be reported as $u(w) = 0.000\ 088$ g.

If the result reported for a measurement represents the average of a number of measurements, rather than a single measurement, the **standard deviation of the mean** ($s/\sqrt{n}$) is the correct estimate of the standard uncertainty. In other words, compared to the uncertainty for a single result, the standard uncertainty for an average result is reduced by a factor of $\sqrt{n}$ (where n is the number of measurement results that were averaged to obtain the reported result).

NOTE: It is important to establish that any software or calculators used to carry out calculations are capable of producing accurate results. For example, some pocket calculators will have difficulty performing the above calculations when using their in-built statistical functions. If you are unable to get the results indicated, you should try coding the data by subtracting 10 and multiplying by 10^4. The transformed data set will then be:

$$1, 0, 2, 2, 1, 0, 1, 0, 2, 0$$

The mean of this transformed set is 0.9 and its standard deviation is 0.875 59. To get the values for the original set, we must reverse the operations performed on it. That is to say, we divide by 10^4 and, in the case of the mean only, we add 10.

As mentioned previously, uncertainty estimates which have not been obtained directly by statistical evaluation of data may need converting into the correct form before the final combination of the individual uncertainty estimates can be carried out. This is discussed in the next section.

Converting Data to Standard Uncertainties Uncertainty estimates can be obtained from a number of sources. However, they may not be expressed as a standard uncertainty and there are a number of rules for converting data.

(i) Data expressed as a confidence interval.

Example: the concentration of a standard solution is stated as 1000 ± 3 mg l^{-1}, where $\pm$ 3 mg l^{-1} represents a 95% confidence interval.

A confidence interval is calculated from $t \times s/\sqrt{n}$ (see Section 6.1.3). To obtain a standard uncertainty we need to calculate $s/\sqrt{n}$. We therefore need to know the appropriate Student t-value (see Appendix, p. 253). However, statements of this type are generally given without specifying the degrees of freedom. Under these circumstances, if it can be assumed that the producer of the material carried out a reasonable number of measurements to determine the stated value, it is acceptable to use the value of t for infinite degrees of freedom, which is **1.96** at the 95% confidence level. If the degrees of freedom are known, then the appropriate t-value can be obtained from statistical tables. In this example, the standard uncertainty is $3/1.96 = 1.53$ mg l^{-1}.

(ii) Data expressed as an expanded uncertainty.

Example: the concentration of a reference solution is $1000 \pm 3\,\mathrm{mg\,l^{-1}}$, where the reported uncertainty is an expanded uncertainty, calculated using a coverage factor of $k = 2$, which gives a level of confidence of approximately 95%.

A coverage factor (usually denoted by the letter k) is used to increase (expand) a standard uncertainty to give the required level of confidence (usually 95%). Expanded uncertainties are discussed in more detail in Section 6.3.6. To convert an expanded uncertainty back to a standard uncertainty, simply divide by the stated coverage factor. In this example, $k = 2$, so the standard uncertainty is $1.5\,\mathrm{mg\,l^{-1}}$.

(iii) Data expressed as a tolerance.

Example: the manufacturing tolerance for a 25 ml Class A pipette is ± 0.03 ml, according to BS 1583 [6]. In this example, a range is given with no indication of the distribution or the level of confidence. We therefore have to make some assumptions about the likely distribution of data within the stated tolerance. We know that there should not be any values outside the tolerance range as this would mean that the equipment was 'out of specification'. We could also assume that values occur anywhere within the tolerance range with equal probability. In the example of a pipette, we are therefore assuming that if we purchased a number of 25 ml pipettes and determined the actual amount of liquid delivered by each pipette (ignoring any random errors), the results would lie anywhere between 24.97 ml and 25.03 ml. There would be no results outside of this range and, if we tested enough pipettes, we would find that the results were evenly distributed across the range. This kind of distribution of data is called a ***rectangular distribution*** and is illustrated in Figure 6.13(a). You can see that the rectangular distribution is very different from the normal distribution (see Section 6.1.2). The normal distribution is unbounded and values near the mean are more likely than values at greater distances from the mean (see Figure 6.4).

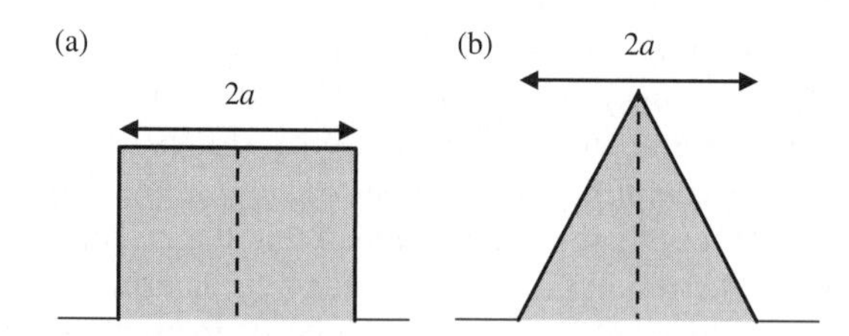

Figure 6.13 (a) Rectangular and (b) triangular distributions of data.

The standard deviation of a rectangular distribution is obtained by dividing the half-range of the distribution (a in Figure 6.13) by $\sqrt{3}$. In the example above, the standard uncertainty is $0.03/\sqrt{3} = 0.017$ ml.

In some cases where a tolerance is quoted, we may have some additional information that leads us to believe that values closer to the centre of the range are more likely than values at the extremes of the range. In such cases, a ***triangular distribution*** is assumed, as shown in Figure 6.13(b). The standard deviation of a triangular distribution is obtained by dividing the half-range, a, of the distribution by $\sqrt{6}$.

It is usually acceptable to assume a rectangular distribution to obtain an initial estimate of the uncertainty associated with manufacturing tolerances. This may, however, overestimate the uncertainty. Some publications assume a triangular distribution for the tolerances associated with volumetric glassware. The rationale for doing this is that a reputable manufacturer is more likely to produce items which have a value closer to the nominal value than at the extremes of the permitted range. In practice, the assumption of a triangular or a rectangular distribution often makes no difference to the overall uncertainty for the results from chemical measurements, as the tolerances associated with volumetric glassware are rarely a significant source of uncertainty.

Table 6.4 summarizes the different rules for converting data to a standard uncertainty.

Table 6.4 Summary of rules for converting data to a standard uncertainty

Data expressed as	Conversion rule
Standard deviation	No conversion required for uncertainty associated with a single value; divide standard deviation by $\sqrt{n}$ for uncertainty associated with the mean of n values
Confidence interval	Divide interval by 1.96 for large number of degrees of freedom (or by appropriate Student t-value if degrees of freedom known)
Expanded uncertainty	Divide by stated coverage factor, k
Stated range (values equally likely across range)	Assume a rectangular distribution; divide half-range by $\sqrt{3}$
Stated range (values close to mean more likely than values at the extremes of the range)	Assume a triangular distribution; divide half-range by $\sqrt{6}$

SAQ 6.4

Convert the following data to standard uncertainties.

(a) The manufacturer's specification for a 100 ml Class A volumetric flask is 100 ± 0.08 ml.

(b) The calibration certificate for a 4-figure balance states that the measurement uncertainty is $\pm$ 0.0004 g, where the reported uncertainty is an expanded uncertainty based on a standard uncertainty multiplied by a coverage factor of $k = 2$, providing a level of confidence of approximately 95%.

(c) The purity of a compound is given by the supplier as 99.9 ± 0.1%.

(d) The standard deviation of repeat weighings of a 0.3 g check weight is 0.000 21 g.

6.3.5.4 Combination

All of the standard uncertainty estimates obtained in the previous stage must now be combined to produce an overall uncertainty. Consider a measurement quantity, y, that is a function of several variables, p, q, ... The model (see Section 6.3.5.1) is $y = f(p, q, \ldots)$. The general expression for combining the standard uncertainties associated with independent variables is as follows:

$$u(y) = \sqrt{\left(\frac{\partial y}{\partial p}\right)^2 u(p)^2 + \left(\frac{\partial y}{\partial q}\right)^2 u(q)^2 + \cdots} \tag{6.11}$$

where $u(y)$, $u(p)$ and $u(q)$ represent the standard uncertainties in the result y and in the parameters p and q, respectively; $(\partial y/\partial p)$ is the partial differential of y with respect to p. The partial differentials are known as ***sensitivity coefficients***. They describe how the result y varies with changes in the parameters p, q, etc.

In many cases the general expression can be reduced to relatively simple expressions for combining uncertainties. The general expression, in equation (6.11), should be used for cases not covered by the equations shown in Table 6.5.

If the model contains a mixture of operations, e.g. $y = (a - b)/(c + d)$, it should be broken down into expressions which consist solely of operations covered by the cases shown in equations (6.12)–(6.15). For example, the expression above should be broken down into the elements $(a - b)$ and $(c + d)$ and the uncertainty for each element calculated by using equation (6.12). The results from these calculations can then be combined by using equation (6.13) to give the combined standard uncertainty.

Table 6.5 Expressions for combining standard uncertainties (c.f. equation (6.11))

Equation for calculating result, y	Combined uncertainty	Equation number
$y = a + b$	$u(y) = \sqrt{u(a)^2 + u(b)^2}$	(6.12)
$y = a - b$	$u(y) = \sqrt{u(a)^2 + u(b)^2}$	
$y = a \times b$	$u(y) = y \times \sqrt{\left(\frac{u(a)}{a}\right)^2 + \left(\frac{u(b)}{b}\right)^2}$	(6.13)
$y = \frac{a}{b}$	$u(y) = y \times \sqrt{\left(\frac{u(a)}{a}\right)^2 + \left(\frac{u(b)}{b}\right)^2}$	
$y = \mathrm{B}x$	$u(y) = \mathrm{B} \times u(x)$	(6.14)[a]
$y = a^n$	$u(y) = y \times n \times \frac{u(a)}{a}$	(6.15)

[a] In equation (6.14), B is a constant with no uncertainty.

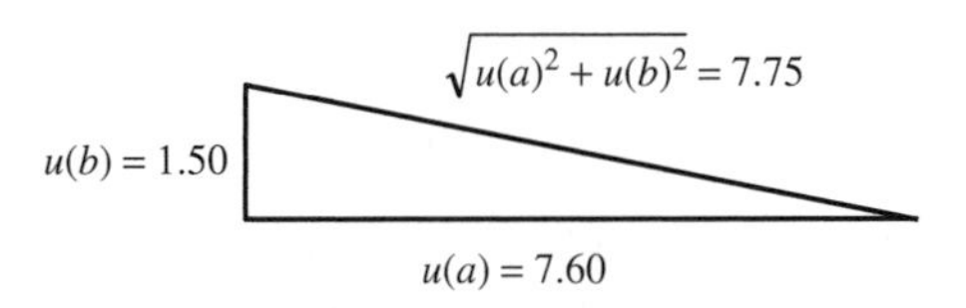

Figure 6.14 Illustration of the combination of standard uncertainties: $u(a)$ is much greater than $u(b)$ and so the combined uncertainty is approximately equal to $u(a)$.

One consequence of the equations for combining standard uncertainties is that the combined standard uncertainty will be dominated by the largest uncertainty components. This is illustrated in Figure 6.14.

It may occur to you that equation (6.15) might be unnecessary since an expression such as $y = a^n$ can be written as $y = aaa\ldots$(i.e. a multiplied by itself n times). This would appear to be a special case of equation (6.13) with b and c equal to a. So, would an application of equation (6.13) give the same result as an application of equation (6.15)?

DQ 6.6

Using $a = 3.72$, $u(a) = 0.19$ and $n = 3$, calculate the uncertainty in y by using equation (6.15) and then equation (6.13).

Answer

Using equation (6.15), the uncertainty is 7.89 while using equation (6.13) it is 4.55. Why do the two equations give different results? The reason is that equation (6.13) is intended for use where the individual variables contributing to a product are **independent**. It may so happen that the values of these variables are the same and possibly (although less likely) the values of their associated uncertainties are the same. Equation (6.15), on the other hand, is intended for those cases where a single variable is multiplied by itself a number of times. In other words, it really is the **same** variable multiplied by itself n times and not a chance equality of intrinsically different variables.

To get a feeling for how these equations operate, let us now put some numbers into them and see what kind of results are produced.

Consider equation (6.12) first. Suppose we have four objects (a, b, c, d) and we wish to know their combined mass (T) and the uncertainty associated with this mass ($u(T)$). The following information is available:

$$a = 27.71 \text{ g}, \;\; u(a) = 0.01 \text{ g}$$

$$b = 32.35 \text{ g}, \;\; u(b) = 0.02 \text{ g}$$

$$c = 47.10 \text{ g}, \;\; u(c) = 0.11 \text{ g}$$

$$d = 19.86 \text{ g}, \;\; u(d) = 0.01 \text{ g}$$

The model for this problem is simply $T = a + b + c + d$.

The combined standard uncertainty associated with T is obtained from:

$$u(T) = \sqrt{u(a)^2 + u(b)^2 + u(c)^2 + u(d)^2}$$

Therefore, $T = 27.71 + 32.35 + 47.10 + 19.86 = 127.02$ g

$$u(T) = \sqrt{0.01^2 + 0.02^2 + 0.11^2 + 0.01^2} = 0.112\ 69 \text{ g}$$

We can now write down the result, which is:

$$T = 127.02\text{g}, u(T) = 0.11 \text{ g}$$

or more compactly:

$$T = (127.02 \pm 0.11) \text{ g}$$

Let us now move on to applications of equation (6.13).

An analyst is preparing a standard solution of concentration C by weighing out a specified amount of a material of known purity and dissolving it in a specified volume of solvent. The following information is available:
mass of material used (M) $= 100.5$ mg, $u(M) = 0.208$ mg
purity of material (P) $= 0.999$ (expressed as a ratio), $u(P) = 0.000\,58$
volume of solvent (V) $= 100$ ml, $u(V) = 0.16$ ml

The model for this problem is:

$$C = \frac{MP}{V} \times 1000 \text{ mg l}^{-1}$$

Note that the factor of 1000 is a conversion factor which is required to obtain the result in the correct units (mg l^{-1}).

The combined standard uncertainty associated with C is obtained from:

$$u(C) = C \times \sqrt{\left(\frac{u(M)}{M}\right)^2 + \left(\frac{u(P)}{P}\right)^2 + \left(\frac{u(V)}{V}\right)^2}$$

Therefore:

$$C = \frac{100.5 \times 0.999}{100} \times 1000 = 1004.0 \text{ mg l}^{-1}$$

$$u(C) = 1004 \times \sqrt{\left(\frac{0.208}{100.5}\right)^2 + \left(\frac{0.000\ 58}{0.999}\right)^2 + \left(\frac{0.16}{100}\right)^2}$$

$$u(C) = 1004 \times \sqrt{0.002\ 07^2 + 0.000\ 581^2 + 0.001\ 60^2}$$

$$u(C) = 1004 \times 0.002\ 68 = 2.69 \text{ mg l}^{-1}$$

The final result can be written as $C = (1004 \pm 2.7)$ mg l^{-1}.

SAQ 6.5

Which of the following equations is the correct one to use to combine standard uncertainties when the measurement model involves only multiplication and/or division?

(a) $u(y) = \sqrt{\left(\frac{u(a)}{a}\right)^2 + \left(\frac{u(b)}{b}\right)^2 + \left(\frac{u(c)}{c}\right)^2}$

(b) $u(y) = \dfrac{y}{\sqrt{\left(\frac{u(a)}{a}\right)^2 + \left(\frac{u(b)}{b}\right)^2 + \left(\frac{u(c)}{c}\right)^2}}$

(c) $u(y) = y \times \sqrt{\left(\frac{u(a)}{a}\right)^2 \times \left(\frac{u(b)}{b}\right)^2 \times \left(\frac{u(c)}{c}\right)^2}$

(d) $u(y) = y \times \sqrt{\left(\frac{u(a)}{a}\right)^2 + \left(\frac{u(b)}{b}\right)^2 + \left(\frac{u(c)}{c}\right)^2}$

(e) $u(y) = y \times \sqrt{\left(\frac{a}{u(a)}\right)^2 + \left(\frac{b}{u(b)}\right)^2 + \left(\frac{c}{u(c)}\right)^2}$

6.3.6 Expanded Uncertainty

In all of the above examples, after combining the standard uncertainties, we have finished by reporting the combined standard uncertainty. We have converted all of our contributory error distributions, be they normal, rectangular or triangular, into equivalent normal distributions and combined them by one means or another. The combined standard uncertainty is equivalent to 1 standard deviation of a normal distribution. You will recall that 1 standard deviation of a normal distribution covers 68.3% of the values in the distribution (see Figure 6.4). By convention, the true value of a measurement result is taken to lie within the uncertainty limits with a probability of 95%. This is approximately equivalent to 2 standard deviations (actually 1.96). In order to bring our reported measurement uncertainty into line with accepted practice, it is necessary to multiply it by 2. The factor 2 is known as a **coverage factor**. We can then add to our report that the confidence level of the quoted uncertainty is approximately 95%. If we wanted to be especially cautious, we could use a coverage factor of 3 to get a confidence level of 99.7%. It was mentioned earlier that standard uncertainties are denoted by the symbol u. Expanded uncertainties are usually denoted by the symbol U.

Consider the previous example of calculating the concentration of a standard solution. The combined standard uncertainty of 2.69 mg l^{-1} would be multiplied by a coverage factor of 2 to give an expanded uncertainty of 5.38 mg l^{-1}. We can now report the result as follows: concentration of solution = (1004 ± 5) mg l^{-1}, where the reported uncertainty is an expanded uncertainty calculated using a coverage factor of 2, which gives a level of confidence of approximately 95%. Note that the coverage factor is applied only to the final combined uncertainty.

It is important to include a statement as to what the value quoted after the $\pm$ represents (i.e. a standard or an expanded uncertainty) so that users of the result interpret the quoted uncertainty correctly.

SAQ 6.6

You have been asked to prepare a 0.1 mol l^{-1} solution of potassium hydrogen phthalate (KHP) for use by colleagues in your laboratory. The solution must be properly labelled and the information you provide on the label must include the concentration of the solution together with a statement of its uncertainty at a 95% confidence level.

The measurement model you will use is:

$$C = \frac{1000 \times M \times P}{V \times F \times 100}$$

where C = concentration of KHP solution (mol l^{-1}), M = mass of KHP taken (g), P = purity of KHP (%), V = final volume of KHP solution (ml) and F = molar mass of KHP (g mol^{-1}).

(continued)

SAQ 6.6 *(continued)*

Calculate the required values on the assumption that you have obtained the following information:

$M = 20.4220$ g, $u_1(M) = 0.000\ 07$ g (balance precision (random effect)), $u_2(M) = 0.000\ 05$ g (balance calibration (systematic effect)).
$P = (99.9 \pm 0.1)\%$ (from supplier's catalogue).
$V = (1000 \pm 0.4)$ ml (from supplier's catalogue), $u_1(V) = 0.10$ ml (standard deviation of replicate measurements of volume of liquid in flask when filled to the calibration mark).
$F = 204.2236$ g mol^{-1}, $u(F) = 0.0017$ g mol^{-1}.

6.3.7 Putting Uncertainty to Use

6.3.7.1 Interpretation of Results

We mentioned at the beginning of this section on uncertainty that every quantitative result we produce should be accompanied by a measure of its quality. The concept of uncertainty has been introduced as a suitable measure of quality, so how then do we put it to practical use?

Each of us is a user as well as a producer of chemical results. Putting on our 'user' hats we will now see how an uncertainty value helps us to interpret an associated chemical measurement.

Suppose we are responsible for accepting or rejecting batches of a certain material used in a manufacturing process. Our decision will be based upon a chemical analysis of the material and one of the criteria for acceptance is that the concentration of compound X in the material shall not exceed a specified level. Given that we have a number of reports in front of us from the laboratory relating to different batches of material, Figure 6.15 shows five possible outcomes. In this figure, the measured concentration of compound X in each of the five cases is shown, together with the expanded uncertainty U ($k = 2$). The reference value that must not be exceeded is also shown. Remember that uncertainty is defined as a range of values within which the quantity being measured is expected to lie. For the result we are considering, this means that the true value could be anywhere in the range $-U$ to $+U$.

In Figure 6.15, only cases (a) and (e) are easy to interpret. Looking at case (a), we see that the measured value is less than the reference value. The upper extreme of the expanded uncertainty is also less than the reference value. We can therefore safely conclude that the concentration of compound X is less than the reference value in case (a) so this particular batch of material can be accepted. In case (e), the measured value exceeds the reference value, as does the lowest extreme of the expanded uncertainty. There is therefore no doubt that the concentration of X exceeds the reference value and the batch of material must therefore be rejected.

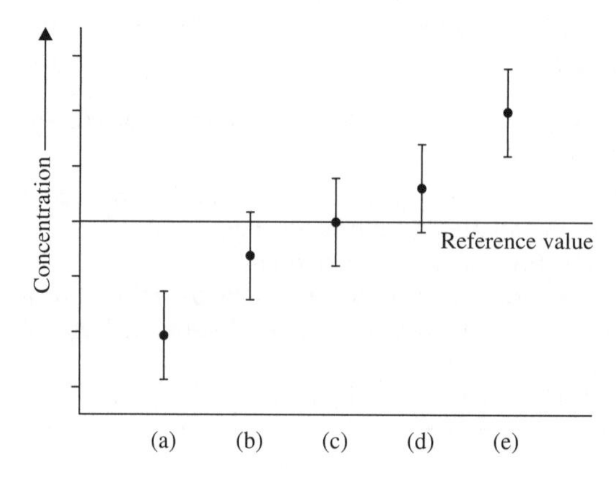

Figure 6.15 Relationship between a reference value and measured values ((a)–(e)) when uncertainty limits are included.

In case (b), although the measured value is less than the reference value, if the expanded uncertainty is taken into account it is possible that the actual concentration of X could exceed the reference value. In case (c), the measurement result equals the reference value. Although the measured value does not exceed the reference value, the expanded uncertainty means that the true value of the concentration of X could exceed the reference value. Finally, in case (d) the measurement value exceeds the reference value but if the expanded uncertainty is taken into account, it is possible that the true value of the concentration of X could be below the reference value.

To decide whether to reject or accept these batches, we would need to know how the reference value was set (e.g. was it set taking into account the possible uncertainty?) and the 'criticality' of the measurement. It is important to discuss such 'borderline' cases with the end-user of the data, and agree how they are to be handled, before the measurements are made.

6.3.7.2 Improving the Quality of Results

A second use for uncertainty values lies in their potential for helping us to improve our experimental procedures. In calculating the uncertainty for a measurement, we will have assembled a list of standard uncertainties for the variables of the measurement model. If we wish to improve the quality of our measurement, we must look first at the component of the measurement system contributing the largest uncertainty. If this is the dominant contribution to the combined uncertainty, then any attempt to improve other aspects of the measurement process will be a waste of time. By attempting to reduce the size of the dominant uncertainty first, we will produce the greatest return for our effort.

Summary

This chapter has considered two key aspects related to quality assurance – the use of control charts and the evaluation of measurement uncertainty. These activities, along with method validation, require some knowledge of basic statistics. The chapter therefore started with an introduction to the most important statistical terms.

The analysis of quality control samples is an important activity for laboratories and to make the most of the data, control charts should be used. This chapter has discussed a number of common types of control chart and described how they are set up and interpreted.

It is important to have some knowledge of the reliability of all measurement results. Measurement uncertainty is the parameter used to describe the range within which the true value (or right answer) for a particular measurement is expected to lie. Evaluating measurement uncertainty involves a number of distinct steps, which are described in this chapter.

References

1. 'Shewhart Control Charts', ISO 8258:1991, International Organization for Standardization (ISO), Geneva, Switzerland, 1991.
2. 'Cumulative Sum Charts – Guidance on Quality Control and Data Analysis Using CUSUM Techniques', ISO/TR 7871:1997, International Organization for Standardization (ISO), Geneva, Switzerland, 1997.
3. 'Guide to Data Analysis and Quality Control Using CUSUM Techniques. Uses and Value of CUSUM Charts in Business, Industry, Commerce and Public Service', BS 5703-1:2003, British Standards Institute (BSI), London, UK, 2003.
4. 'Guide to Data Analysis and Quality Control Using CUSUM Techniques. Introduction to Decision-Making Using CUSUM Techniques', BS 5703-2:2003, British Standards Institute (BSI), London, UK, 2003.
5. 'Guide to Data Analysis and Quality Control Using CUSUM Techniques. CUSUM Methods for Process/Quality Control Using Measured Data', BS 5703-3:2003, British Standards Institute (BSI), London, UK, 2003.
6. 'Specification for One-Mark Pipettes', BS 1583:1986, British Standards Institute (BSI), London, UK, 1986.
7. 'Guide to the Expression of Uncertainty in Measurement', ISO/IEC Guide 98:1995, International Organization for Standardization (ISO), Geneva, Switzerland, 1995.

Chapter 7

Benchmarking Your Laboratory

Learning Objectives

- To understand the difference between proficiency testing schemes and collaborative studies.
- To know the key features in the organization of each of the above activities.
- To be familiar with the different approaches to obtaining the assigned value and target range for a proficiency testing scheme.
- To understand how z-scores are calculated and interpreted.

The previous chapters of this book have discussed the many activities which laboratories undertake to help ensure the quality of the analytical results that are produced. There are many aspects of quality assurance and quality control that analysts carry out on a day-to-day basis to help them produce reliable results. Control charts are used to monitor method performance and identify when problems have arisen, and Certified Reference Materials are used to evaluate any bias in the results produced. These activities are sometimes referred to as internal quality control (IQC). In addition to all of these activities, it is extremely useful for laboratories to obtain an independent check of their performance and to be able to compare their performance with that of other laboratories carrying out similar types of analyses. This is achieved by taking part in interlaboratory studies. There are two main types of interlaboratory studies, namely ***proficiency testing (PT) schemes*** and ***collaborative studies*** (also known as collaborative trials).

Quality Assurance in Analytical Chemistry E. Prichard and V. Barwick

DQ 7.1

What do you think is the main difference between a proficiency testing scheme and a collaborative study?

Answer

A proficiency testing scheme tests the performance of the participating laboratories whereas a collaborative study is used to test the performance of a particular method.

These two different types of interlaboratory study are discussed in detail in the following sections.

7.1 Proficiency Testing Schemes

The primary aim of a proficiency testing scheme is to provide laboratories with a framework for obtaining a regular independent assessment of their performance. A key feature of proficiency testing schemes is that the assessment of laboratory performance is expressed in terms of a score that can be readily interpreted in terms of statistics. The scoring system used must be applicable to a variety of situations. In particular, it must be possible to apply it to a range of analyte concentrations. For example, in some schemes participants will be asked to determine the analyte at a range of concentrations. The acceptable standard deviation of results at a relatively low concentration (e.g. 1 mg kg^{-1}) may be different from the acceptable standard deviation at a higher concentration (e.g. 500 mg kg^{-1}). The scoring schemes commonly used in proficiency testing schemes are discussed in Section 7.3.3.

There are two main types of proficiency testing scheme. First, there are those set up to assess the competence of a group of laboratories to undertake a very specific analysis, e.g. lead in blood or the number of asbestos fibres in air collected on membrane filters. Secondly, there are those schemes used to evaluate the performance of laboratories across a certain sector for a particular type of analysis. Because of the wide range of possible analyte/matrix combinations it is not practicable to assess the performance of laboratories when analysing all the possible sample types. Instead, a representative cross-section of analyses is chosen (e.g. determination of different pesticide residues in a range of foodstuffs or the determination of trace levels of metals in water samples).

Each of these two main types of proficiency testing schemes can be further subdivided into three categories, as follows:

(a) Randomly selected subsamples from a bulk homogeneous supply of material are distributed simultaneously to participating laboratories, as shown in

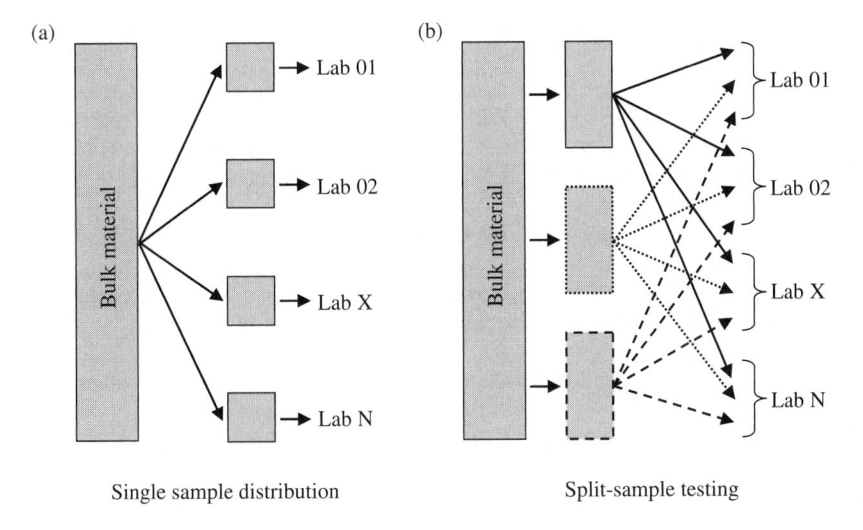

Figure 7.1 Sample distribution schemes for proficiency testing schemes: (a) single sample distribution; (b) split-sample testing.

Figure 7.1(a). This is by far the most common type of proficiency testing scheme.

(b) Samples of a product or a material are divided into two or more parts; each participating laboratory tests a subsample of each part. This is frequently referred to as 'split-sample' testing and is illustrated in Figure 7.1(b). This type of scheme generally involves comparing data produced by a small number of laboratories, often as a means of evaluating them as potential or continuing suppliers of particular analytical services.

(c) The sample to be tested is circulated successively from one laboratory to the next. In some instances, the sample is returned to a central laboratory before being passed onto the next testing laboratory in order to determine whether any changes to the sample have taken place. This type of scheme is sometimes referred to as a *measurement comparison scheme*. Note that such schemes are uncommon in chemical testing.

In some sectors, particularly clinical analysis, proficiency testing is referred to as External Quality Assessment (EQA).

There are numerous proficiency testing schemes available, operated by a number of different organizations, Table 7.1 gives some examples. Further information on the range of proficiency testing schemes available can be found on EPTIS, a web-based Proficiency Testing information system. This is a database of PT schemes covering most measurement sectors (www.eptis.bam.de).

Table 7.1 Examples of proficiency testing schemes organized by UK providers

Scheme	Provider	Scope
Aquacheck	LGC	Clean and waste waters, sludges, sediments and soil
CONTEST (Contaminated Land Proficiency Testing Scheme)	LGC	Contaminated soils
EQA for Food Microbiology	Health Protection Agency	Food borne micro-organisms
FAPAS (Food Analysis Performance Assessment Scheme)	Central Science Laboratory	Proximates, trace contaminants in food
LEAP (Laboratory Environmental Analysis Proficiency)	Central Science Laboratory	Water, effluent, contaminated land
MAPS (Malt Analytes Proficiency Testing Scheme)	LGC	Malt and barley
QMS (Quality Management's Quality in Microbiology Scheme)	LGC	Microbiological examination of food and food ingredients
QWAS (Quality Management's Quality in Water Analysis Scheme)	LGC	Microbiological assessment of waters, effluents and sludges
RICE (Regular Interlaboratory Counting Scheme)	Health and Safety Laboratory	Asbestos fibre counting in the construction industry
TOYTEST	LGC	Toy safety
WASP (Workplace Analysis Scheme for Proficiency)	Health and Safety Laboratory	Hazardous airborne substances

7.2 Organization of Proficiency Testing Schemes

Irrespective of the type of proficiency testing scheme, it is usually organized in a sequence of clear steps.

The scheme organizer lays down the rules for the conduct of the tests and the interpretation of the data. This is circulated to the participants so that they understand exactly how the scheme is run and how their results are assessed. The organizer should establish an expert advisory group which will address issues relating to the operation of the scheme, such as the tests and types of samples to be tested under the scheme, how the performance of participants is to be judged, technical difficulties raised by participants and the performance of participants in the scheme as a whole.

Materials are chosen such that they are, as far as possible, representative of the type of material that is normally analysed. This will be in terms of the matrix and the concentration range of the analyte. Materials must be tested for **homogeneity** before distribution since the effective interpretation of all of the test data for the scheme is based on this assumption – each separate batch of test material must be checked and this is generally carried out by a single expert laboratory. Precision is the important aspect as the aim is for all subsamples to have as near as possible the same composition. Non-homogeneity is always possible after the material has been distributed, due to sedimentation and separation. Therefore, if a participant is to analyse a subsample of the material supplied by the organizers, it is important to re-homogenize the whole sample before taking a portion of the material for analysis. It is also important that the samples remain stable during the period of the proficiency testing round.

There is no experimentally established optimum **frequency** for the distribution of samples. The minimum frequency is about four rounds per year. Tests that are less frequent than this are probably ineffective in reinforcing the need for maintaining quality standards or for following up marginally poor performance. A frequency of one round per month for any particular type of analysis is the maximum that is likely to be effective. Postal circulation of samples and results would usually impose a minimum of two weeks for a round to be completed and it is possible that over-frequent rounds have the effect of discouraging some laboratories from conducting their own routine quality control. The cost of proficiency testing schemes in terms of analysts' time, cost of materials and interruptions to other work has also to be considered.

Once the samples have been analysed, the results are reported by the laboratory to the scheme organizer who produces a statistical score for each laboratory (see Section 7.3.3). The participants are informed of the outcome as soon as possible after the closing date for the reporting of results so that they can respond to any problems. Usually the results are in the form of a report, which includes detailed information on the participants' performance, the overall distribution of results and so on. A key aspect of proficiency testing schemes is their confidentiality. Typically, each participating laboratory is given an identification number/code, known only to themselves and the scheme organizers. Reports from the scheme identify the participants only by their identification numbers. Each laboratory can therefore identify their own performance and compare it with the performance of other laboratories, but without knowing their identities.

In the early rounds of proficiency testing schemes, there is usually an overall significant improvement in performance. However, for some laboratories there can be a lack of consistency – achieving a good performance in one round, but not being able to sustain it over a long period – suggesting that they do not have an adequate quality management system in place.

The operation of proficiency testing schemes is described in the harmonized protocol produced by IUPAC [1]. Further information can be found in ISO/IEC Guide 43 [2] and ILAC Guide 13 [3].

7.3 The Statistics Used in Proficiency Testing Schemes

As mentioned previously, one of the key aspects of proficiency testing is the provision of a 'score' which participants can use to evaluate their performance. There are a number of different scoring systems. The majority involve comparing the difference between a laboratory's result (x) and some target or assigned value (X), with a target range such as a standard deviation ($\hat{\sigma}$). Each scoring system has some acceptability criteria, so that participants can evaluate their performance. This section explains how assigned values and target ranges are obtained, and describes the scoring systems commonly used. A detailed discussion of the statistics used in proficiency testing schemes is contained in ISO 13528:2005 [4].

All proficiency testing schemes should have a statistical protocol which states clearly how the data will be processed and how laboratory performance will be evaluated. This protocol should also describe how the assigned value for any parameter in a test sample is estimated. This is an important consideration, as the performance of individual laboratories is gauged by comparison with the assigned value.

7.3.1 The Assigned Value

The assigned value is the value attributed to the quantity being measured in the proficiency testing exercise. It can be considered the 'target value' for participants. There are a number of different approaches to determining the assigned value. These are by use of:

- formulation;
- certified reference values;
- reference values;
- consensus values from expert laboratories;
- consensus values from the scheme participants.

The approach is chosen by the scheme organizer, in consultation with the members of the scheme. Each approach has advantages and disadvantages, as discussed in the following sections.

7.3.1.1 Formulation

In this context, formulation involves the preparation of 'synthetic' test materials. A known amount or concentration of the analyte is added to a suitable material (i.e. the sample matrix) containing none of the analyte of interest (or to a material containing a very small but well characterized amount of the analyte). Such test materials are also known as 'fortified' materials. The formulation method is particularly useful when it is the amount of analyte added to particular test objects (e.g. air filters) that is being determined. One advantage of this approach is that the analyte can usually be added to the test material in very accurate amounts using gravimetric or volumetric methods. It is therefore usually relatively straightforward to establish both the traceability and the uncertainty of the assigned value.

However, there are a number of problems associated with this approach. It may be difficult to produce material of sufficient homogeneity owing to problems in mixing the analyte with the sample matrix. It may also be difficult to obtain a suitable blank or sufficiently well characterized test material. In addition, a 'synthetic' material may well behave differently from a 'natural' material. The analyte may be in a different chemical form or less strongly bound to the sample matrix. For these reasons, the formulation approach is not appropriate for many sample types.

7.3.1.2 Certified Reference Values

Provided the sample matrix and analyte concentration are appropriate, matrix Certified Reference Materials (CRMs) can make ideal proficiency testing samples. The assigned value is the certified value given on the certificate accompanying the CRM. The certificate will also give an uncertainty estimate for the certified value, and the use of CRMs allows the traceability of analytical data to be established. However, matrix CRM availability is limited and the materials are often expensive. Hence, Certified Reference Materials are seldom used as PT samples.

7.3.1.3 Reference Values

An alternative to using matrix CRMs as the proficiency test material is for a single expert laboratory to determine the assigned value, using either a primary method (see Chapter 5, Section 5.2) or a fully validated method calibrated using CRMs. This effectively produces a reference material for distribution as the proficiency testing material. However, a laboratory able to carry out a suitable primary method may not be available for all sample types. An alternative approach is for a single laboratory to analyse both the test material and appropriate CRMs, using a suitable method under 'repeatability conditions'. In effect, the method is calibrated using the CRMs, providing direct traceability and an uncertainty for the test material. The lack of suitable CRMs is a disadvantage of this approach.

7.3.1.4 Consensus Values from Expert Laboratories

In this case, the assigned value is obtained from data produced by a number of **expert laboratories** who have analysed the proficiency testing sample by using a recognized reference method. The laboratories must be able to demonstrate their proficiency in the measurement of interest, be accredited, use CRMs and validated methods and have excellent performance in the proficiency testing scheme. The measurements may be carried out prior to the distribution of the test sample or as part of the proficiency testing round. The main disadvantage of this approach is that the expert laboratories are not infallible and there may be an unknown bias in the results that they produce. However, compared to using a single laboratory the risk of the assigned value being biased is reduced, as the probability of all of the expert laboratories producing incorrect results is relatively low. Another potential problem with this approach is that choosing the expert laboratories may be difficult and controversial.

7.3.1.5 Consensus Values from the Scheme Participants

Frequently, the assigned value is taken as the consensus of the results from all of the participants in the proficiency testing round. However, if the consensus was simply taken as the mean of all of the results this may not give a suitable estimate of the assigned value: the mean will be influenced by any results which are significantly different from the majority of the results (sometimes referred to as **outliers**). A method of reducing the influence of any outliers on the assigned value is therefore needed. The assigned value is sometimes taken as the average of the participants' results, after the removal of any outliers (outlier tests, such as the Dixon Q test or Grubbs' tests, are used to check for outliers [5]). However, outlier tests are seldom completely satisfactory, especially when there are many possible outliers in a data set. A more modern approach is to apply *robust statistics*. When using robust statistics, the influence of extreme points in a data set is reduced, therefore giving a better estimate of parameters such as the mean and standard deviation. One advantage of this approach is that it removes the problem of identifying possible outliers and deciding what to do with them. There are many types of robust statistics. Some common approaches are discussed in Section 7.3.6.

A consensus value is the easiest and cheapest method for obtaining the assigned value. It is the appropriate approach when all participants are using a single standardized empirical method and a large number of laboratories are involved.

As the consensus approach is straightforward, it is frequently used for non-empirical methods, particularly when the other approaches to obtaining the assigned value are not feasible. However, there are a number of disadvantages and potential problems associated with this approach. First, it is not uncommon to find that there is no real consensus within a group of laboratories. Secondly, the group consensus may be significantly biased. For example, problems with the analysis of tin in tomato paste have arisen which led to a consensus assigned value significantly lower than the actual amount of tin present in the sample [6]. The use of a consensus value in such circumstances may perpetuate poor methodology (over time, the participants may gain closer agreement with each other, but they will be converging on the wrong answer!). Finally, it is difficult to establish the traceability of the assigned value unless the methodologies used by each of the participating laboratories are known in detail.

7.3.2 The Target Range

The target range is the standard deviation used in the assessment of proficiency. It is set by the scheme organizer, usually with specialist advice. The target range is intended to represent the acceptable variability for a particular analysis. The value is chosen so that an acceptable performance score represents results that are fit for a particular purpose. Ideally, the same target range – either in absolute terms (e.g. $1\,\text{mg kg}^{-1}$) or proportional to the assigned value (e.g. 10% of the assigned value) – should be used over successive rounds of the proficiency test so that performance scores are comparable over time. As the target range directly influences the performance score awarded to a participant, changing the range round-on-round will make it difficult for participants to assess real changes in performance.

As in the case of setting the assigned value, there are a number of different approaches to defining the target range:

- prescribed value;
- by perception;
- results from a collaborative study;
- general model;
- data obtained from the proficiency testing round.

7.3.2.1 Prescribed Value

In this case, the target range is chosen to ensure that laboratories obtaining a satisfactory score are producing results that are fit for a particular purpose. The chosen value is therefore related directly to a 'fitness for purpose' statement, which may be derived from a legislative requirement.

7.3.2.2 By Perception

The target range can also be set at a value that corresponds to the level of performance that the scheme organizer, and the participants in the scheme, would like laboratories to be able to achieve. This is sometimes referred to as setting the target range 'by perception' and is equivalent to using a 'fitness for purpose' statement.

It is not always easy or possible to set a target range based on 'fitness for purpose' criteria and so in some cases alternative approaches are used.

7.3.2.3 Results from a Collaborative Study

If the analytical method used by participants in the proficiency testing round has been validated by means of a formal collaborative trial, then the repeatability and reproducibility data from the trial can be used. The repeatability standard deviation gives an estimate of the expected variation in replicate results obtained in a single laboratory over a short period of time (with each result produced by the same analyst). The reproducibility standard deviation gives an estimate of the expected variation in replicate results obtained in different laboratories (see Chapter 4, Section 4.3.3 for further explanation of these terms).

If σ_R represents the reproducibility standard deviation and σ_r represents the repeatability standard deviation, the 'between-laboratory' standard deviation, σ_L, is calculated from the following:

$$\sigma_L = \sqrt{\sigma_R^2 - \sigma_r^2} \quad (7.1)$$

The target range for proficiency assessment, $\hat{\sigma}$, is then calculated from:

$$\hat{\sigma} = \sqrt{\sigma_L^2 + \frac{\sigma_r^2}{n}} \quad (7.2)$$

where n is the number of replicate measurements each participant has to perform on a test sample in a round of the scheme.

7.3.2.4 General Model

If data from a collaborative study are not available to estimate the reproducibility of a method, a general model such as the Horwitz function can be used. The Horwitz function is described in Chapter 4, Sections 4.4 and 4.6.2. It can be used to predict the value of σ_R based on the concentration of the analyte in the proficiency test material. The disadvantage of this approach is that the chosen model may not accurately represent the true reproducibility of the method.

7.3.2.5 Data Obtained from Proficiency Testing Round

The final approach to establishing the target range is to use the participants' data from the proficiency testing round. To avoid problems associated with identifying and rejecting possible outliers, a robust estimate of the standard deviation is preferred (see Section 7.3.6). The main disadvantage of this approach is that the target range may vary substantially from round to round, hence making it difficult for participants to look for trends in their results over time.

7.3.3 Performance Measures

There are a number of different performance scoring systems used in proficiency testing schemes. The most commonly used system is the ***z*-score**, which is calculated from the following:

$$z = \frac{(x - X)}{\hat{\sigma}} \tag{7.3}$$

where x is the result submitted by the laboratory, X is the assigned value and $\hat{\sigma}$ is the target range. The **z-score** is based on the properties of the normal distribution. In a normal distribution of data, 95% of values are expected to lie within ± 2 standard deviations of the mean (see Chapter 6, Section 6.1.2 for further information on the normal distribution). Hence a score of $|z| \leq 2$ is considered a **satisfactory** result. A z-score between 2 and 3 is considered a **questionable** result, as there is a 5% chance that a reported result that is 2 standard deviations removed from the true value is actually drawn from a population whose mean is in fact the same as the true value (i.e. there is actually nothing wrong with the result). However, there is only a very small chance (approximately 0.3%) that a result that is more than 3 standard deviations away from the mean is actually drawn from a population whose mean is the same as the true value. A score of $|z| \geq 3$ is therefore considered an **unsatisfactory** performance. An example of the calculation of z-scores is given in Section 7.3.6.

The z-score takes no account of the uncertainties in the assigned value or the participants' results. The **z'-score** is similar to the z-score, but takes into account the uncertainty in the assigned value, u_X, as shown in the following equation:

$$z' = \frac{(x - X)}{\sqrt{\hat{\sigma}^2 + u_X^2}} \tag{7.4}$$

It is interpreted in the same way as a z-score, with the same values used to indicate questionable and unsatisfactory performance. The z-score and z'-score are related as shown in the following equation:

$$z' = z \times \frac{\hat{\sigma}}{\sqrt{\hat{\sigma}^2 + u_X^2}} \tag{7.5}$$

Therefore, z'-scores will only differ significantly from z-scores if there is significant uncertainty in the assigned value.

The z'-score does not take into account the uncertainty in the participant's result. The ***zeta*-score** is another variation on the z-score. It takes into account both the uncertainty in the assigned value, u_X, and the uncertainty in the participant's result, u_x:

$$zeta = \frac{(x - X)}{\sqrt{u_x^2 + u_X^2}} \tag{7.6}$$

This approach requires a valid estimate of the uncertainty from each participant. Where there is an effective system in operation for validating participants' estimates of the uncertainty in their results, *zeta*-scores can be used instead of z-scores and interpreted in the same way.

There are a number of other scoring systems but these are not as widely used as the z-score system. **E_n numbers** take into account the expanded uncertainty in the assigned value (U_{ref}) and the expanded uncertainty in the participant's result (U_x):

$$E_n = \frac{x - X}{\sqrt{U_x^2 + U_{ref}^2}} \tag{7.7}$$

E_n numbers are used when the assigned value has been produced by a reference laboratory, which has provided an estimate of the expanded uncertainty. This scoring method also requires a valid estimate of the expanded uncertainty for each participant's result. A score of $|E_n| < 1$ is considered satisfactory. The acceptability criterion is different from that used for z-, z'- or *zeta*-scores as E_n numbers are calculated using expanded uncertainties. However, the E_n number is equal to *zeta*/2 if a coverage factor of 2 is used to calculate the expanded uncertainties (see Chapter 6, Section 6.3.6). E_n numbers are not normally used by proficiency testing scheme providers but are often used in calibration studies.

Laboratory biases can be expressed in absolute terms, as a ***Q*-score**, or as a percentage (***D*-score**):

$$Q = \frac{(x - X)}{X} \tag{7.8}$$

$$D_{\%} = \frac{100(x - X)}{X} \tag{7.9}$$

The D-value is interpreted in the same way as z-scores, for example, a score of $|D_{\%}| \leq 200\hat{\sigma}/X$ indicates satisfactory performance.

The overall distribution of Q-scores is expected to be centred on zero. This will be the case if:

- participants' results are used to estimate the assigned value (assuming a small number of outliers);
- the assigned value is obtained from a group of expert laboratories and the other participants in the round do not show a bias relative to the expert laboratories;
- the assigned value is obtained by formulation and there is no widespread use of methods which lead to biased results.

Q-scores are less commonly used in the UK than z-scores, and tend to be applied in schemes where some form of data transformation is carried out prior to calculating the scores.

7.3.4 Combination of z-Scores

There are a number of situations where the combination of several z-scores to produce a single statistic might appear useful. For example, in a single proficiency testing round a number of tests may be carried out, each producing a z-score. Participants may want a single score that summarizes their overall performance in that round. It may also be considered useful to summarize performance for a particular test over a number of rounds. There are various methods of combining z-scores. One approach is the **rescaled sum of z-scores** (RSZ), given by:

$$RSZ = \sum_{i=1}^{n} \frac{z_i}{\sqrt{n}} \tag{7.10}$$

where n is the number of scores to be combined. The RSZ can be interpreted in the same way as a z-score.

Another approach is the **sum of squared z-scores** (SSZ):

$$SSZ = \sum_{i=1}^{n} z_i^2 \tag{7.11}$$

If such combined scores are used, great care must be exercised to avoid incorrect conclusions or misleading statements. The RSZ value will tend to hide a small proportion of moderately high z-scores among mostly acceptable scores, as high scores of opposite sign will cancel out. The SSZ is sensitive to single outliers. For these reasons, the use of combination scores is not generally recommended and is, in fact, strongly discouraged in ISO 13528 [4]. A simple graph plotting the variation in z-scores with time is more useful.

7.3.5 Interpretation of Performance Scores

In Section 7.3.3, we learned about one of the most common scoring systems used in proficiency testing – the z-score – and how laboratory performance can be judged as being satisfactory, questionable or unsatisfactory. How should a laboratory act on the performance scores it receives? Action should be considered in the following situations:

- when an unsatisfactory result has been obtained (this is mandatory for laboratories accredited to ISO/IEC 17025);
- when two consecutive questionable results have been obtained for the same test method;
- when nine consecutive results with the same bias against the assigned value, for the same test method, have been obtained.

It is always useful to consider the performance in a proficiency testing round in a wider context. One of the main factors to consider is the performance of all of the participants in the round. If the majority of the results are satisfactory, but yours is not, this is likely to indicate a problem in your laboratory. However, it is worth remembering that your laboratory may have got the correct result and the other participants are in error! In addition, if many other participants also have unsatisfactory results, there is still a problem, but it is less likely to be in your laboratory.

In some cases, unsatisfactory performance may be due to the test method used by the laboratory being inappropriate or having poorer performance characteristics (e.g. precision and bias) than methods used by other participants. If the proficiency testing scheme organizer has set the target range, $\hat{\sigma}$, based on a standard method with superior performance characteristics, then lesser methods will be more likely to result in unsatisfactory performance scores. In such a situation you should try to compare your performance with that of other laboratories that are using the same analytical method, if this information is available in the proficiency testing report. Ideally, proficiency testing samples should always be similar to those routinely analysed by participants in their laboratories. However, proficiency testing schemes frequently cover a range of sample types and so for any given laboratory some samples may be unusual or extreme in their composition or nature. If your laboratory's performance is unsatisfactory for a particular sample, it is worth considering whether the sample was within the normal scope of operations. Unusual sample matrices can cause problems with extraction of the analyte and sample clean-up. Analyte concentrations much higher or lower than those found in routine test samples may also cause problems.

Although the providers of proficiency testing schemes should have a quality management system in place, on occasions problems can arise which will affect the quality of the data evaluation being carried out. These can include transcription errors during data entry, mistakes in the report, software problems and inappropriate criteria for evaluation being used. Such problems should be remedied by the provider once the problem has been identified.

Ultimately, it is up to the laboratory to carry out a thorough investigation when unsatisfactory scores are received. If all of the possibilities discussed above have been considered and ruled out as major contributions to the poor performance, the laboratory needs to look at its own procedures in some detail. There are many possible causes of unsatisfactory performance. They can generally be subdivided into two categories – analytical errors and non-analytical errors. Both are equally important.

DQ 7.2

What do you think are common causes of analytical and non-analytical errors?

Answer

There are many possible sources of error which can lead to unsatisfactory performance. Some of the common ones are listed below.
Sources of *analytical errors* include the following:

- calibration of equipment;
- problems with instrumentation;
- problems with extraction and clean-up of the sample;
- sample outside of method scope (e.g. different matrix or analyte concentration);
- interferences;
- analyst error such as incorrect preparation of calibration standards or incorrect dilution of samples.

Non-analytical errors include the following:

- calculation errors;
- transcription errors;
- reporting results in the wrong units.

7.3.6 Robust Statistics

As mentioned in Sections 7.3.1.5 and 7.3.2.5, participants' data may be used to calculate the assigned value and the target range. In any set of proficiency testing data, it is likely that there will be a number of extreme values or outliers. When using the standard statistical techniques described in Chapter 6, Section 6.1.3, these outliers can have a significant effect on estimates of the mean and standard deviation of the data. This could ultimately lead to misleading performance scores being generated for that round of the scheme. When processing the data, something has to be done to minimize the effect of any extreme values. The traditional approach was to use statistical tests to confirm the presence of outliers, eliminate the extreme values identified and recalculate the mean and standard deviation. This is not always a satisfactory approach as there is an element of judgement involved in identifying outliers and deciding whether or not they should be rejected.

In recent years, the use of *robust statistics* has become the favoured approach for obtaining sound estimates of the average value and spread of a data set. The advantage of robust statistics is that no rejection of suspect data is required.

The simplest estimate of a robust average is the **median** of the data – when data points are arranged in order of magnitude, the median is the middle value of the series. A robust standard deviation is frequently obtained from the median of all of the absolute differences from the sample median (usually abbreviated as MAD, median absolute deviation). This value is multiplied by 1.483 to give a figure equivalent to a normal standard deviation. This gives the MAD_E value, which is an estimate of the robust standard deviation.

Table 7.2 and Figure 7.2 show data obtained from one round of a proficiency scheme for the determination of the alcoholic strength of a spirit. The results are expressed as % alcohol by volume (%abv). When the participants' results are ranked in order of magnitude, the median is the middle value, which in this case is 40.04. The assigned value is therefore 40.04 %abv.

To calculate the robust standard deviation for this data set, you first have to calculate the absolute difference between each result and the median, $|x_i - \text{median}|$, and then find the median of these values. The median absolute deviation (MAD) is 0.02 %abv. This is converted to a standard deviation equivalent (MAD_E) by multiplying by 1.483:

$$\text{Robust standard deviation, } MAD_E = 1.483 \times 0.02 = 0.03 \text{ \%abv.}$$

For comparison, the mean of the data set is 40.02 %abv and the standard deviation is 0.05 %abv. You can see that the robust standard deviation is substantially smaller than the standard deviation. The use of robust statistics has reduced the influence of the extreme values in the data set.

The target range for this proficiency testing scheme has been set at 0.03 %abv, based on 'fitness for purpose' criteria. Using an assigned value (X) of 40.04 %abv

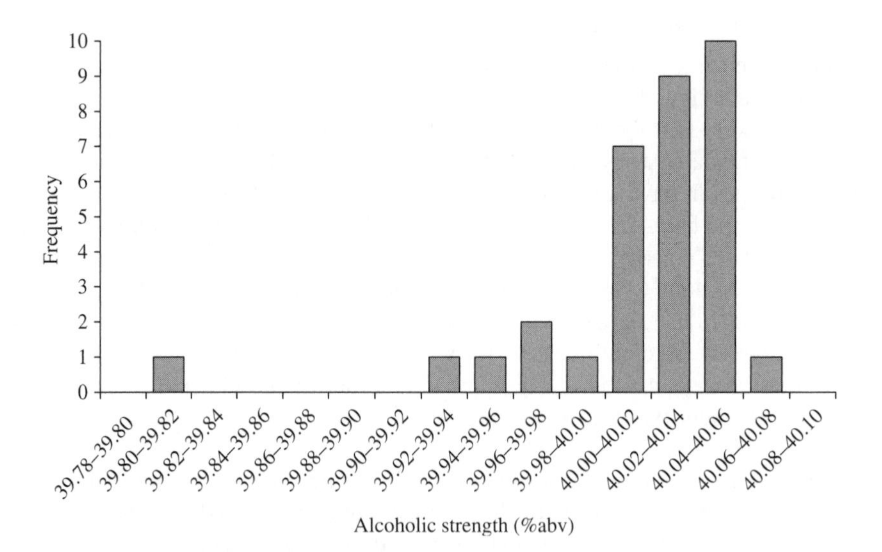

Figure 7.2 Histogram of data from a proficiency testing round for the determination of the alcoholic strength of a spirit.

Table 7.2 Data obtained from a proficiency testing round for the determination of the alcoholic strength of a spirit (median = 40.04 %abv, $\hat{\sigma} = 0.03$ %abv)

Laboratory identity number	Result (%abv)	$\|x_i - \text{median}\|$	z-score
1	40.04	0.00	0.0
2	40.02	0.02	−0.7
6	39.81	0.23	**−7.7**
7	40.04	0.00	0.0
8	40.05	0.01	0.3
9	40.02	0.02	−0.7
12	40.04	0.00	0.0
13	40.02	0.02	−0.7
14	40.05	0.01	0.3
16	40.00	0.04	−1.3
18	39.93	0.11	**−3.7**
19	40.05	0.01	0.3
20	40.06	0.02	0.7
21	40.03	0.01	−0.3
22	39.98	0.06	−2.0
24	40.05	0.01	0.3
27	40.04	0.00	0.0
28	40.02	0.02	−0.7
29	40.02	0.02	−0.7
31	40.05	0.01	0.3
32	40.04	0.00	0.0
35	40.07	0.03	1.0
42	40.01	0.03	−1.0
47	40.06	0.02	0.7
49	39.98	0.06	−2.0
50	40.04	0.00	0.0
52	39.95	0.09	**−3.0**
57	40.03	0.01	−0.3
60	40.01	0.03	−1.0
62	40.05	0.01	0.3
64	40.06	0.02	0.7
68	40.05	0.01	0.3
78	40.04	0.00	0.0

and the above target range we can now calculate the z-scores, using equation (7.3); the results are shown in Table 7.2. Figure 7.3 shows a plot of the z-scores. In this example, three laboratories had unsatisfactory performance ($|z| \geq 3$). The performance of all the other laboratories is considered acceptable as their z-scores are $\leq |2|$.

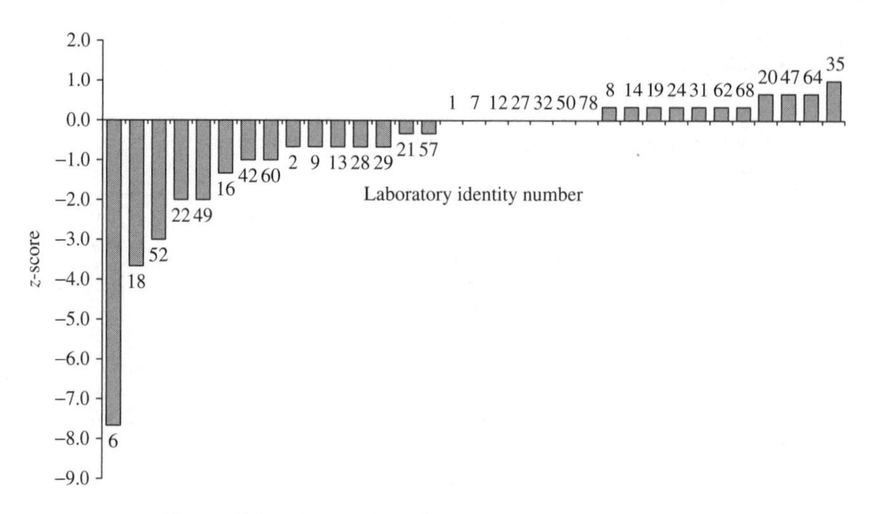

Figure 7.3 z-Score plot of the data given in Table 7.2.

7.4 Making the Most of Participation in Proficiency Testing Schemes

Participation in proficiency testing schemes can bring significant benefits to laboratories. However, the proficiency testing scheme itself cannot cause improvements in laboratory performance. It is up to the participants to use the feedback they receive from the scheme to monitor their performance and to implement improvements where necessary.

Ideally, laboratories should view participation in proficiency testing schemes as an educational activity. It is important to learn from the experience, regardless of whether the performance has been satisfactory or unsatisfactory. Both positive and negative feedback are valuable. It is important to understand how to get the optimum benefit from participation in proficiency testing. Laboratories should treat the proficiency testing samples in the same way as routine samples. If you are involved in proficiency testing, it will generally be obvious when you are analysing the proficiency testing sample as opposed to a routine test sample. However, you should try to treat the sample as you would a normal test sample. There is no benefit for the laboratory if proficiency testing samples are treated in any special way. The laboratory will learn very little about the quality of its routine work if the proficiency testing samples are not handled in the same way as a routine test sample.

Participation in proficiency testing schemes is an ongoing activity. It is therefore useful to monitor performance over a period of time and to look for trends. Performance over time can be demonstrated statistically by using measures such as *RSZ* and *SSZ* (see Section 7.3.4) but as mentioned previously, these can be misleading. It is better to monitor performance scores by plotting them on a chart. Charts can show trends in performance and are particularly valuable in highlighting potential measurement problems before they become serious. This approach can also be used to show whether an individual unsatisfactory result is a 'one-off' or part of a more serious longer-term problem.

In addition to monitoring performance over time, performance for groups of related analytes can be studied. In many proficiency testing schemes, there are groups of common analytes which are measured using the same analytical method. An example of this is the determination of trace metals, which are extracted from the matrix together, and measured in a single analysis by inductively coupled plasma–optical emission spectroscopy (ICP–OES). Looking at results for such groups can show whether any analytical problems are generic (i.e. all results are unsatisfactory) or specific to one or more analytes (i.e. most, but not all, results are satisfactory). Where all results in the group are unsatisfactory, the problem is almost certainly generic and only one investigation and corrective action is necessary.

SAQ 7.1

Table 7.3 Data obtained from a proficiency testing round for the determination of moisture in barley

Laboratory identity number	Moisture (wt%)
1	13.4
2	13.5
3	13.4
4	13.2
5	13.6
6	12.7
7	13.3
8	13.6
9	13.6
10	13.4
11	13.2
12	13.7
13	13.4
14	13.3
15	13.7
16	13.2
17	13.3

The data shown in Table 7.3 are from one round of a proficiency testing scheme for the determination of moisture in barley. There are a number of different

(*continued overleaf*)

SAQ 7.1 *(continued)*

methods that can be used for determining moisture content. The laboratories reporting results in Table 7.3 all used an oven-drying method.

(a) Calculate the mean, standard deviation, robust average (median) and robust standard deviation (MAD_E) of the data (if you don't know how to calculate the mean and standard deviation, see Chapter 6, Section 6.1.3).

(b) The scheme organizers, in consultation with the scheme participants, have set a target standard deviation ($\hat{\sigma}$) of 0.2 wt%. Choose an appropriate assigned value and calculate a *z*-score for each of the laboratories.

(c) How many of the laboratories would have their performance judged as satisfactory, questionable and unsatisfactory?

7.5 Collaborative Studies

As mentioned previously, a *collaborative study* is a test of an analytical method rather than of laboratory performance. Each laboratory uses a defined method to analyse identical portions of homogeneous materials. It is then possible to assess the performance characteristics of that method of analysis. Collaborative studies may be used to develop a standard method of analysis. Governments, trade associations or standards organizations (e.g. ISO) may require a standard method to be established for a particular analyte in a given matrix. A working group of experts in this area of analysis will be set up and a list prepared of laboratories that will participate in the study. The working group will appoint a co-ordinator and the collaborative study will then be organized. The typical sequence of events for such an exercise is shown below.

(i) The text of a proposed method is sent to all of the participants.

(ii) Comments regarding the method are sent back to the co-ordinator.

(iii) The revised text of the method and the samples are sent to the participants.

(iv) Participants analyse the samples.

(v) The results are sent to the co-ordinator for statistical analysis.

(vi) A report of the study is sent to the participants.

(vii) A proposal for a method is made by the co-ordinator, in consultation with the participants.

The aim of the collaborative study is to evaluate the performance of the method (in particular its precision, see Chapter 4, Section 4.3.3) and establish whether it is 'fit for purpose'. Unlike proficiency testing, a collaborative study is typically a one-off event although steps (ii) to (vi) may have to be repeated before a satisfactory method can be agreed. Another distinction between proficiency tests and collaborative studies is the nature of the laboratories that participate. Proficiency testing schemes are generally open to all laboratories, whereas participants in collaborative studies are pre-selected on the basis of their track record in a particular type of analysis, or have to demonstrate their competence as part of the study through the analysis of standards and quality control samples.

The ISO 5725 series of standards describes in detail the organization of a collaborative study [7–12]. Further information can be found in Horwitz [13].

Summary

This chapter has considered two of the types of interlaboratory comparison exercise in which your laboratory may participate. It is important to remember that proficiency testing schemes and collaborative studies have different aims. The former is a test of the performance of the laboratory, whereas the latter is used to evaluate the performance of a particular analytical method. Laboratories should participate in proficiency testing schemes (where an appropriate scheme is available) as this provides an independent check of the laboratory's performance. This chapter has described the key features of proficiency testing schemes and explained how the results from participation in a scheme should be interpreted.

References

1. Thompson, M., Ellison, S. L. R. and Wood, R., *Pure Appl. Chem.*, **78**, 145–196 (2006).
2. 'Proficiency Testing by Interlaboratory Comparisons – Part 1: Development and Operation of Proficiency Testing Schemes', ISO/IEC Guide 43-1:1997, International Organization for Standardization (ISO)/International Electrotechnical Commission (IEC), Geneva, Switzerland, 1997.
3. 'Guidelines for the Requirements for the Competence of Providers of Proficiency Testing Schemes', ILAC G13:2007, International Laboratory Accreditation Cooperation (ILAC), Silverwater, Australia, 2007.
4. 'Statistical Methods for Use in Proficiency Testing by Interlaboratory Comparisons', ISO 13528:2005, International Organization for Standardization (ISO), Geneva, Switzerland, 2005.
5. Miller, J. N. and Miller, J. M., *Statistics and Chemometrics for Analytical Chemistry*, 5th Edition, ISBN 0-13-129192-0, Pearson Education Ltd, Harlow, UK, 2005.
6. Sargent, M. and Holcombe, G., *VAM Bull.*, Issue 34, 19–23 (2006). [http://www.nmschembio.org.uk] (accessed 8 November, 2007).
7. 'Accuracy (Trueness and Precision) of Measurement Methods and Results – Part 1: General Principles and Definitions', ISO 5725-1:1994, International Organization for Standardization (ISO), Geneva, Switzerland, 1994.
8. 'Accuracy (Trueness and Precision) of Measurement Methods and Results – Part 2: Basic Method for the Determination of Repeatability and Reproducibility of a Standard Measurement Method', ISO 5725-2:1994, International Organization for Standardization (ISO), Geneva, Switzerland, 1994.

9. 'Accuracy (Trueness and Precision) of Measurement Methods and Results – Part 3: Intermediate Measures of the Precision of a Standard Measurement Method', ISO 5725-3:1994, International Organization for Standardization (ISO), Geneva, Switzerland, 1994.
10. 'Accuracy (Trueness and Precision) of Measurement Methods and Results – Part 4: Basic Methods for the Determination of the Trueness of a Standard Measurement Method', ISO 5725-4:1994, International Organization for Standardization (ISO), Geneva, Switzerland, 1994.
11. 'Accuracy (Trueness and Precision) of Measurement Methods and Results – Part 5: Alternative Methods for the Determination of the Precision of a Standard Measurement Method', ISO 5725-5:1998, International Organization for Standardization (ISO), Geneva, Switzerland, 1998.
12. 'Accuracy (Trueness and Precision) of Measurement Methods and Results – Part 6: Use in Practice of Accuracy Values', ISO 5725-6:1994, International Organization for Standardization (ISO), Geneva, Switzerland, 1994.
13. Horwitz, W., *Pure Appl. Chem.*, **67**, 331–343 (1995).

Chapter 8

Documentation and its Management

Learning Objectives

- To understand the principles involved in good record keeping.
- To have a basic knowledge of the factors to include in controlled documents.
- To appreciate good practice in generating reports for customers.
- To describe what is meant by opinions and interpretation.
- To appreciate what is involved if opinions and interpretation form part of the scope of accreditation.

8.1 Documentation

The top-level document in a management system is usually called a Quality Manual. This is a generic document that sets out the structure of the management system and the management's policy on key aspects of the system. The details on how the management system is operated and the procedures that staff have to follow to ensure quality results/services are set out in a range of supporting documentation. The interrelationship between the documents is shown in Chapter 9, Figure 9.1.

8.1.1 Quality Manual

The content, size and format of the Quality Manual will depend on the nature of the organization. As described in Chapter 9, the quality management standards

Quality Assurance in Analytical Chemistry E. Prichard and V. Barwick

used by laboratories, ISO/IEC 17025 and ISO 9001 allow some flexibility about its content [1,2]. When a laboratory decides to compile its Quality Manual, the process of agreeing exactly what the management system is and preparing the manual may well bring to light a number of inconsistencies. It may reveal that there are different practices and maybe some activities are not being carried out. The Quality Manual should cover activities that are general throughout the organization. The content will include sections such as:

- a quality policy statement;
- the scope of the management system;
- the general organization and management structure;
- the roles and responsibilities of the Technical Management and the Quality Manager;
- a quality audit and review programme;
- a description of the interrelationship between the supporting documentation;
- contract negotiations and customer requirements;
- how to deal with customer complaints;
- document control;
- staff competence and training;
- the laboratory environment;
- technical requirements of the quality management system;
- site security.

The manual sets out the policy in each area, not the detailed instructions as to how procedures are carried out.

8.1.2 Supporting Documentation

The supporting documentation sets out how the management system is operated and the procedures that members of staff have to follow to ensure the quality of results and services. The relationship between the documents is shown in Chapter 9, Figure 9.1. Quality Procedures (QPs) are often held centrally within the organization so that they are available to all members of staff, whereas the Standard Operating Procedures (SOPs) and Work Instructions (WIs) are held locally as they only apply to specific areas of the organization. Some examples of what might be classed as a QP are details of document control, audit and review procedures, general aspects of contract review, preventive and corrective actions, method validation, etc.

DQ 8.1

What do you think should be included in a SOP or a WI?

Answer

Note that some organizations may not use the terminology used in this book and may not distinguish between SOPs and WIs. Standard Operating Procedures provide details of how a series of operations are carried out. An example of a SOP would be the detailed instruction for carrying out a particular analytical method. Work Instructions give details of how a specific operation is carried out. What might be classed as a WI is how to operate a particular instrument, how to estimate measurement uncertainty or how to calibrate a piece of equipment.

You may have included some other documents with the SOPs and WIs because they are also held in your area of the laboratory. These could be, for example, guides for carrying out particular activities, such as method validation, and international guidelines on how to achieve reliable results. Some areas of activity may be covered by legislation; copies of the relevant documents will be kept locally as it may be referenced in the SOP. Equipment manuals are also kept locally.

8.1.3 Record Management

Documentation is an important aspect of any laboratory management system and there are several types of documentation. It will include all the procedures, e.g. organization charts, Standard Operating Procedures, instrument registers, daily check-logs and instructions and records relating to particular samples or a particular contract. Such documentation affects every aspect of the laboratory's operations by showing what has happened previously, what is happening at present and what is expected to happen in the future. Together, they provide a record of events. A good system for record keeping is an essential ingredient in any well run laboratory and provides the basis for an effective quality system. This is managed by having a record management system. This will show how records are generated, used, changed, preserved and controlled.

8.1.4 Records

A 'record' is a piece of information permanently or semi-permanently preserved on a particular medium. A laboratory will generate a large number of records in various forms (such as paper, photographic film, computer files, video or audio tape). For the purposes of this book, the spoken word does not constitute a record unless it is recorded in some way. These records encapsulate the what, how, when, why and who of any activity. In the following sections, discussions focus on the use of paper records (still the prevalent type). The same principles, however, are applicable to all other types of record.

The purpose of a record is to enable undistorted retrieval of the information as and when required. A record needs to be such that someone in the future can follow through exactly what was done. It is essential that records are complete and that they are retained for an adequate period of time. Records are used in laboratories for a number of reasons, e.g. monitoring, controlling, communicating and proof.

DQ 8.2

Consider the four examples of typical laboratory activities listed below and in each case try to list some examples of records which might be found for each activity:

(a) purchasing;

(b) laboratory procedures;

(c) laboratory/customer dialogue;

(d) analytical work.

Answer

Records that might be found for each application include the following:

(a) Purchasing – purchase orders, invoices, receipts, inventories, tenders, financial accounts.

(b) Laboratory procedures – analytical methods, rules for calibrating instruments, maintenance and cleaning procedures, training records, procedures for recording customer complaints, quality control.

(c) Laboratory/customer dialogue – requests for analysis, cost estimates, work orders, analytical reports, invoices.

(d) Analytical work – workbooks and sheets, analytical data, quality control charts, calibration records.

One thing that is apparent is that there are a wide variety of types of record put to many different uses. In order to ensure that records fulfil their purpose effectively, a system for generating, identifying, controlling, copying, using, removing from circulation, changing, storing and archiving records is required. In an age where word processing is widely used, setting up such a system is now comparatively straightforward.

8.1.5 Generating Records

A laboratory needs to ensure that for each type of record it defines rules governing the right and wrong ways to produce it. This is necessary to achieve consistency and to ensure the record always contains the appropriate level of information, presented in a way that is readily understood. This preferred way of producing records is often characterized as a 'format' or 'layout'. If you were to ask a number of people to document a particular procedure, you might well be quite alarmed at the variety of descriptions you would get. This is because different people have quite different views on what they would consider important and on what they would take for granted. It is quite a challenge to write good clear procedures or records that almost anyone else could use. Training and practice are usually needed before being able to do it well. Working to pre-agreed layouts or formats helps to simplify the process.

Getting the content of a record right is important and frequently quite difficult. It may be appropriate to limit the generation of particular types of record to particular 'expert' people. For example, analytical methods and procedures need to be written-up so that the contents are easy to follow and leave no opportunity for ambiguity. The instructions conveyed by these procedures must be safe, unambiguous and sufficiently detailed to ensure that whoever uses the procedure can understand what he/she is meant to do.

A number of individuals may be responsible for producing a particular type of record, for example, documenting methods. In order to ensure consistency, it is normal to have one person with overall authority to issue the records as fit for use.

For recording results, observations and other analytical data, it may be convenient for the laboratory to design, produce and use pre-printed forms. This has a number of advantages. It ensures uniformity in the way results are recorded and calculations are made, it facilitates staff training, it simplifies checking and detecting errors and maintains cost-effective consistency of approach.

8.1.6 Record Identification

An important part of any record management system is the ability to take a record, and recognize what it is, what it contains, who produced it and when, whether the person producing it was authorized to do so, whether the contents are still current, its confidentiality status, its copying status and whether it is complete. Much of this can be achieved using simple identifiers on each page of the document. The remainder can be achieved using inventories and lists. Record inventories can be used to list the history of records, which version is current, and which staff are authorized to produce or amend particular types of record, etc.

Every page of a record should have the following information:

- title (usually full title on first page and an abbreviated form on successive pages);

- version number;
- date;
- page number (of total number – often ignored, but IMPORTANT);
- security status.

In addition the front page should have details of:

- copying restrictions;
- author and authorizer;
- issuing person;
- distribution list;
- copy number (of total copies).

These details easily can be incorporated by using word processing.

8.1.7 Document and Record Control

Documents are only of use if everyone has complete confidence in them. The QPs, SOPs and WIs that are issued, as part of the management system, must be reviewed and approved by authorized personnel. The laboratory has to establish a document control system. Included in this will be details of the way that documents should be uniquely identified (issue date, issue number, page number and total number of pages), the frequency of review of documents to ensure continued suitability and the method by which obsolete documents are removed from circulation. If changes to a document are required, these have to be approved. The laboratory document control system can allow for amendments to be made by hand. Each amendment needs to be clearly marked, initialled and dated. There should be an indication of how many changes can be made this way. There will be cases where the level of complexity of the change is such that an amendment by hand is no longer permissible and a new version has to be issued. Details are required to identify a 'complex change'. In all cases, a new version should be introduced as soon as possible. Each of the documents is numbered so that it is possible to track it down to the person who issued it. All other documents, such as Standards, regulations, drawings, software, instruction manuals and specifications, also need to be controlled. Some records and procedures will be kept in a computerized system. There have to be procedures in place to allow for changes to computerized records and information about how these processes are controlled.

DQ 8.3

Why do you think records need to be controlled?

Answer

Records describe what has been done and the outcome of the activity. It is important that records are kept safely and that unrecorded changes are not made. Therefore, records have to be controlled in much the same way as documents. It is important that records are clearly identified, accessible and retained for as long as the customer requires. This can be up to 30 years but could be much shorter. The archiving of records can be in any format. However, it is important that they do not deteriorate and that there is a mechanism for reading them if they are in electronic format.

The management system has to contain procedures for dealing with **complaints**. This will generate yet another form of record. The complaint has to be recorded and the action(s) taken to resolve it. There will be corrective actions and preventive measures put in place as a result of the complaint.

This reinforces the lessons learnt in the previous section. The system for using various records should be defined in supplementary instructions, which should be enforced as necessary. The user is responsible for ensuring that the record in use is the appropriate version. Only those persons with the necessary authority should amend text, or remove or change records, and when doing so should ensure that changes are brought to the notice of those using the records. Records frequently contain confidential information, which should be highlighted by the appropriate page identifiers. Users are responsible for ensuring this confidentiality is not compromised, e.g. sensitive documents should not be left lying around for unauthorized people to read.

8.1.8 Reporting Results

Once analytical results have been produced, invariably a certain amount of manipulation is necessary to translate the results into information that can be understood by the customer. The reporting analyst may have to sort and process a large and varied amount of information in order to produce a small number of final answers. Data from standards may be used to produce calibration curves or calibrate instrument response. Results from quality control samples will have been plotted on charts to ensure that the system was working satisfactorily at the time the measurements were made. Sample data will be quantified by comparison with the standards and suitable corrections made. Then, checks may be made to confirm the results by examining the answers to look for any obvious wrong data. It is appropriate for someone independent to check at least some of the data transcriptions and calculations. Finally, results should be expressed to the correct number of significant figures or decimal places and declared with the appropriate degree of uncertainty.

It is unfortunate, especially where great care has been taken in gathering the data, if mistakes are then made during the reporting process which renders the effort wasted. The calculation and reporting processes are made easier if the data are recorded in a clear manner in the first place. The need for checking data from first principles cannot be over emphasized. Once the final answers have been obtained and any additional conclusions or opinions reached, a report can be compiled from the information for communication to the customer.

The essence of good reporting is to provide the information clearly and unambiguously in a form that suits the customer. An obvious requirement therefore is to recognize the customer's needs. Customers of analytical services have a wide variety of backgrounds. On the one hand, a customer might have no scientific background but submit a sample for analysis to find out whether or not it conforms to a particular specification. In such a case, the answer required of the analyst is a simple 'yes' or 'no'. At the other extreme, the customer may be another analyst with full understanding of the background of the required test, but without the necessary resources to carry out the test themselves. In this case, the customer may require copies of all of the data generated from the test so that the original calculations may be checked or new calculations made. Between these two extremes there will clearly be other types of customer, who have a little or a lot of knowledge of the science behind the tests and the corresponding requirements which they want to have reported to them. In each case, where a job is agreed with a customer, the level of information to be reported should be agreed beforehand. If no such format has been agreed, then care should be taken not to 'pad out' the report with unnecessary information that may confuse the customer. Give the customer the information necessary to answer the immediate problem, but make it clear to the customer that additional information is available if required.

Where a laboratory is working to a particular quality standard, there may be particular requirements governing the level of information to be included in a report to a customer. In cases where such a level of information might confuse the customer, it is normally possible, with the agreement of the customer and the body overseeing the quality standard, to obtain dispensation to provide a simplified report, provided that the omitted information is available and can be reported if required.

In general an analytical report should be compiled using some or all of the following information:

- detail of analysing laboratory;
- unique report reference;

- customer details;
- date(s) of receipt of samples;
- sample details, including reference numbers, descriptions, amount and condition received;
- date of analysis of samples;
- reference to tests carried out (including overview of the principles of the method);
- details of special conditions;
- analytical results, including measurement uncertainty;
- evidence of metrological traceability of results;
- limits of detection;
- recovery data (if required by the customer);
- precision data (if required by the customer);
- conclusions and recommendations;
- disposal details;
- name and signature of analyst issuing the report and date of issue.

8.1.9 Copying Records

Perhaps this section should be called restrictions on copying! When individuals create copies of records, without having the authority to do so, it is a 'recipe for chaos'. This is particularly so when key records are subject to regular update. Consider the following example.

Ten individually numbered copies of a method are used in a laboratory and kept centrally in a drawer. An analyst makes several unauthorized copies of the method for his/her own convenience. The method becomes obsolete and an updated version is reissued, the obsolete official versions are collected from the drawer and ten copies of the updated version are issued in their place. This all takes place while the analyst with the photocopies is away on holiday. The photocopies of the obsolete version are not withdrawn because only the analyst knows about them. On return from holiday, the analyst continues to use the obsolete version, blissfully ignorant of the update. Two different versions of the method are now in use.

Various lessons can be learnt from this example. Apart from the need to have copying restrictions on the document, there is a need to have enforced instructions to ensure the restrictions are adhered to. Similarly, it may be appropriate to restrict access to copying facilities.

8.1.10 Storing and Archiving Records

Storage facilities for records should reflect the need to preserve confidentiality, integrity and logical retrieval. Thought should be given to the susceptibility of the records to damage from fire (and heat), flood (and humidity), electric or magnetic fields, dust, solvents, sunlight or other radiation.

8.2 Opinions and Interpretations

Some customers will require the results of the analysis to be interpreted or an opinion given, e.g. does the discharged water from the factory comply with the current legislation(?) or how many cigarettes can be made from 2 kg of cannabis leaf? There is now provision for this aspect of the work of an analyst to be included in the scope of accreditation of the laboratory to the Standard ISO/IEC 17025:2005 [1]. It is important to be clear about what is meant by *Opinions* and *Interpretations*. In the context of this book, it is the subjective expression given that is based on results, academic or scientific knowledge and experience gained over a period of time.

8.2.1 Examples where Opinions and Interpretations may be Requested

It is a common misconception that opinions and interpretations are only offered by forensic scientists and Public Analysts. Analysts from many areas are required to provide this service, e.g. those dealing with consumer safety, geology/geochemistry, oil exploration and food science, to mention but a few. Some examples are given below.

A building consultant has asked a laboratory to carry out a series of measurements on a range of determinands in a number of soil samples taken from an area of land. The purpose of the analysis is to determine if the soil is contaminated. Based on the results, it is the opinion of the laboratory that the land is contaminated and that the soil needs to be removed to a depth of 2 m and filled with 'clean' soil.

Another case might be an in-house investigation of a manufacturing plant effluent. Samples are sent to a laboratory and the results from a number of analytical tests indicate that there has been an effluent system failure. In the opinion of the laboratory, the likely cause of the failure is blocked filters.

Opinions may also include how a result can be used in the legal defence or prosecution of an individual or organization. This can arise from a number of different cases, such as alcohol levels in a blood sample, contamination of a foodstuff, meat content of pies or whether oil comes from a particular oil field.

When opinions and/or interpretations of results are required, it is essential that this is established during the setting-up of the contract to carry out the work. If this is not done thoroughly, then the appropriate procedures may not be in place to ensure the opinion and/or interpretation is based on objective evidence. It may be necessary to have a chain of custody of the samples, or statistical analysis of the results may be required before a conclusion can be reached. When statistical analysis is required, it may be important, e.g. that the appropriate number of results are gathered or that particular checks are in place. A robust contract review process is an essential element of a laboratory's management system if it wishes to demonstrate its competence to provide an opinion or an interpretation.

8.2.2 Accreditation of Opinions and Interpretations

As mentioned previously, it is possible to include opinions and interpretations within the scope of accreditation. It should be realized that the opinions and interpretations themselves are not accredited. This accreditation is only given if the work is already accredited to ISO/IEC 17025 [1]. What is required by the Standard is evidence of the procedure used by the laboratory to authorize an individual to give an opinion or interpretation. This means there has to be a procedure within the management system of the laboratory that sets out the criteria upon which the quality of the person giving the opinion is assessed. This may be split into two parts, namely the criteria for assessing competence and the criteria for assessing experience.

DQ 8.4

Suggest the criteria you would use to assess competence and those you should use to assess experience.

Answer

Some of the criteria you might have considered are shown below but there may be others appropriate for your laboratory.

Some of the criteria for assessing the **competence** of an analyst include:

- expertise and relevant knowledge;
- professional judgement;
- professional integrity;
- due diligence;
- impartiality;
- presentation skills;
- training received.

Some criteria for assessing the **experience** of an analyst include:

- length of service;
- numbers of relevant samples analysed;
- knowledge of sample type;
- number of times in court as an expert witness;
- number of peer reviewed papers;
- professional recognition.

Your list may be different but so long as it can be justified it is equally valid.

For a large laboratory, it may be convenient to tabulate the individuals who are authorized to give an opinion or interpretation against the tasks for which it is relevant.

Accredited laboratories using the United Kingdom Accreditation Service (UKAS) logo on their reports should make it clear to their customers whether the opinion and/or interpretation is part of their scope of accreditation. If the opinion or interpretation on a report is not within the scope, then there has to be a disclaimer, 'The opinions and interpretations indicated are outside the scope of UKAS accreditation'. UKAS has published a guide to help interpret the ISO standard [3].

Summary

This chapter covers the different types of documentation found in an analytical laboratory. These include the documents which are part of the management system and those dealing with the activities in specific areas of the laboratory. The control of documents is also covered. There are sections on the production and management of records. Finally, there is a brief description of what is meant by the accreditation of opinions and interpretations.

References

1. 'General Requirements for the Competence of Testing and Calibration Laboratories', ISO/IEC 17025:2005, International Organization for Standardization(ISO)/International Electrotechnical Commission (IEC), Geneva, Switzerland, 2005.
2. 'Quality Management Systems – Requirements', ISO 9001:2000, International Organization for Standardization (ISO), Geneva, Switzerland, 2000.
3. 'UKAS Guidance on the Application of ISO/IEC 17025 Dealing with Expressions of Opinions and Interpretations', LAB13, United Kingdom Accreditation Service (UKAS), Feltham, UK, 2001.

Chapter 9
Managing Quality

Learning Objectives

- To realize the benefits to a laboratory of a quality management system.
- To understand how a laboratory selects a particular Standard as being suitable for demonstrating the quality of their work.
- To be able to identify the major components which are required in a laboratory's quality management system.
- To understand the difference between auditing a quality management system and conducting a quality system review.
- To be able to plan an audit.
- To appreciate the inputs necessary for a laboratory to conduct a quality system review.
- To be able to define the responsibilities of staff at all levels towards laboratory quality.

In the previous chapters of this book, we have looked at many aspects of quality in laboratories. Some of the relevant Standards have been mentioned and their similarities and differences outlined. This chapter aims to give more detail on the components of the Standards and show how a quality management system can be achieved in the laboratory. The documentation required and the processes necessary to demonstrate that the management system operates to the requirements of International Standards will be explained. It is important to be clear that the overall management system of a laboratory or organization will cover all of their operations; this includes quality, administration and technical systems.

A laboratory's management system is a system that establishes policy and objectives and how to achieve those objectives. A component of this will be the management system covering all aspects of quality in the laboratory. In many ways, a system is really just common sense procedures adopted by a laboratory, written down on paper, to ensure consistency of application.

9.1 The Management System

Over a period of time, any operating laboratory will develop a range of procedures to help it carry out its work. Some of the laboratory's procedures will be written down while others will be known by particular individuals in the laboratory. Some will be considered to be general knowledge – or at least everyone will think that they are until the day someone does something wrong because they were not aware of the correct procedures! The details of a laboratory's quality management system therefore need to be written down so that everyone in the laboratory can see what the system is and what is expected of them. As mentioned in Chapter 2, the main component of this documentation is usually referred to as the Quality Manual. In most laboratories, the Quality Manual sets out the structure of the quality management system and will be supported by a whole range of other more detailed documentation, such as calibration records, for example.

When a laboratory decides to produce its Quality Manual, the process of agreeing the requirements of a quality management system and what to include in the manual may bring to light a number of inconsistencies, e.g. differences in practice and opinion within the laboratory. There may be other unexpected discoveries, e.g. that some activities are not being carried out at all. Once all of these issues have been argued out and the agreed procedures have been written down, the staff of the laboratory will have a reference book – the Quality Manual – to which they can refer if they need to check on how something should be done. In addition there will be a number of other documents setting out the detailed procedures.

DQ 9.1

List the work instructions and standard procedures used in your laboratory that you think contribute to your laboratory's quality management system and should therefore appear in a Quality Manual.

Answer

Your list will probably include the names of those who are responsible for various aspects of the work, what happens to samples and in what sequence, how equipment is calibrated and operated, how samples are labelled and where they are kept and what checks are made to ensure that results are valid. Many other procedures, such as who is responsible for stocking the stationery cupboard or whose turn it is to make the tea, are all vital to the smooth operation of the laboratory, but do not directly affect the quality of the laboratory's work. They are therefore not part of the quality management system and do not really belong in the Quality Manual.

The content of a Quality Manual for a reasonably large laboratory might be along the lines of the list given in Table 9.1.

9.1.1 The Benefits of a Management System

There will be costs in adopting a quality management system but there are compensating benefits. In an established business consisting of many laboratories, each operating complex procedures, the more likely it is that misunderstandings and mistakes will have crept in and been adopted even though they are bad practice. Even in small laboratories, the absence of a member of staff who is on holiday or ill can cause confusion. If operating procedures are written down for staff to refer to, as part of the quality management system, the number of such mistakes will be reduced.

A laboratory that discovers that it has issued incorrect results faces the daunting prospect of informing its customers of what has happened and offering to re-analyse the relevant samples. It may also face demands for damages to compensate customers for costs which have arisen as a result of actions taken based on the laboratory's erroneous results. Ensuring that the normal operating systems minimize the mistakes that are made means that the number of occasions when extra work is required, to put things right, after an error has occurred is minimized. This results in significant cost-savings to the laboratory.

This concept of 'getting it right first time' is now being adopted throughout industry, and anyone who has experienced problems with faulty products, from a motor car to a toaster, will appreciate the value of this approach in improving the customer's perception of the company involved. Any organization which develops a proven reputation for producing reliable products has a major advantage over rival organizations whose products, however impressive their advertised performance, are considered to be unreliable. Therefore, a further positive benefit for laboratories is that the customers of laboratories are increasingly asking for evidence that the laboratory's results are reliable. The easiest way for customers to do this is to insist that an appropriate independent accreditation body accredits any laboratory tendering for their business. When a laboratory has established its quality management system and has had this assessed and accredited by an external accreditation body, the laboratory can use this recognition of their standards as a positive advertisement for their services. In fact, increasing numbers of the customers of laboratories are insisting that any laboratory undertaking work for

Table 9.1 Example of a contents list of a Quality Manual

Table of contents of a Quality Manual	
Section 1	*General*
	1 Terms and definitions
	2 Introduction and objectives
	3 Vision
	4 Quality policy
	5 Activities and scope
Section 2	*Management system requirements*
	6 General requirements, hierarchy and organization
	7 Planning
	8 Document and record control
	9 Staffing
	9.1 Competence, awareness and training
	10 Accommodation, environment and security
	11 Corporate support services
	11.1 Sample reception and post-room
	11.2 Stores
	11.3 Information technology
	12 Tenders and contracts
	12.1 Contract review
	13 Purchasing
	13.1 Purchasing process
	13.2 Subcontractors
	14 Communication with customers
	15 Marketing, communications and information
	15.1 Responding to the media
	15.2 Marketing
	16 Cause analysis, corrective and preventive action
	17 Audit and management review
Section 3	*Technical requirements*
	18 Traceability of physical and chemical measurements
	18.1 Uncertainty of measurement
	18.2 Sample identification and data traceability
	18.3 Selection and validation of methods
	18.4 Infrequently used methods
	18.5 Validation of instrumental software
	18.6 Validation of spreadsheets
	18.7 Participation in Proficiency Testing (PT) schemes/interlaboratory comparisons and review of data
	18.8 Use of quality control materials and reference materials
	18.9 Use of quality control charts
	18.10 System suitability checks

Table 9.1 (*continued*)

Table of contents of a Quality Manual
19 Handling and storage of samples and materials
19.1 Customer samples
19.2 Good Laboratory Practice
19.3 Operation and management of PT schemes
20 Equipment
20.1 Control of monitoring and measuring equipment
20.2 Calibration certificates
20.3 Calibration of equipment and volumetric glassware
20.4 Dealing with errors
21 Sampling
22 Reporting of results
23 Measurement, analysis and improvement process
24 Continual improvement system
25 Opinions and interpretations
26 Statistical techniques

them must have a quality management system in place that has been recognized as meeting agreed International Standards. Customers are only too aware of the risks of having work carried out by 'cut-price' operators and are increasingly seeking reputable laboratories to undertake their work.

The process of formalizing the laboratory's procedures into a quality management system and documenting the system in a Quality Manual is rather like trying to run a comb through a tangled mass of hair. The end result is a uniform, structured approach to laboratory quality which is easier to manage and much more pleasing to the eye of the beholder. However, the process of untangling all of the threads is time-consuming (and can be 'painful').

The initial resource requirement will therefore be the effort required to agree the quality management system structure and content and compile the information as a Quality Manual. This process often takes several months to complete and, according to the size of the laboratory concerned, may involve a number of staff allocating a significant proportion of their time to this work. It must be remembered that the quality management system must be tailored to the work – not the work tailored to the quality management system.

9.1.2 Types of Management Standards for Laboratories

Different individuals and different laboratories can have very different views of which quality issues are important and what standards need to be set. This could lead to each of the customers of laboratories having to check that each laboratory that they send work to meets the standard of quality that they require. This would waste a great deal of time and provoke endless disagreements between customers and laboratories over which quality measures are or are not necessary. As mentioned in Chapter 2, a number of organizations have already developed and published Standards for quality management systems which are relevant for laboratories. These Standards can therefore be 'taken off the shelf' and referred to by laboratories and by their customers.

The two principal organizations that have prepared and published Standards for quality management systems that are relevant to analytical chemistry laboratories are:

- the Organization for Economic Co-operation and Development (OECD), which has developed the Good Laboratory Practice principles, often referred to as 'GLP';
- the International Organization for Standardization (ISO), which has produced a range of standards and guidance relevant to laboratories.

In addition, there are national bodies that produce specific Standards for their own country. Where there is a comparable ISO Standard, the National Standard is just a change of numbering, e.g. in the UK the ISO 9001:2000 standard becomes BS EN ISO 9001:2000.

The Standards that are most relevant to laboratories have already been mentioned in Chapter 2. The management staff of a laboratory will look at all possibilities, decide on the Standard which best suits their organization and then design a quality management system to meet the chosen Standard's requirements. One has to remember that in terms of the management standard, quality means fitness for purpose. Senior staff in an organization will decide on the management standard that is most appropriate for their business.

DQ 9.2

Write down what you think should be taken into consideration when choosing an appropriate Standard for the laboratory.

Answer

A number of things will be taken into account. The choice will be based on the size and nature of the business, e.g. is analysis performed to support a manufacturing process or is the business providing a testing and calibration service for external customers? If a laboratory needs its quality assessed for more purposes than just reporting data, i.e. it needs also to cover processes and decision making, ISO 9001:2000 is appropriate [1]. If some of the work of an organization is involved with registration of new regulated substances, then GLP will be required for that work [2]. Laboratories who are mainly concerned with demonstrating their competence in testing or calibration will select ISO/IEC

17025:2005 [3]. Larger businesses that carry out a number of functions, in addition to calibration and testing, may require certification to ISO 9001:2000, as well as accreditation to ISO/IEC 17025:2005. Clinical laboratories may select either ISO 15189:2007 [4] or ISO/IEC 17025:2005.

Potential customers can be confident that a laboratory which meets the requirements of the appropriate Standard will also satisfy their requirements without having to undertake their own inspections. This is why organizations are stipulating that laboratories must meet the requirements of one of the internationally recognized Standards before they can be considered for contracts to carry out analytical work.

SAQ 9.1

You are employed in an analytical laboratory in a group measuring the concentration of residues of pesticides. Your company is developing tests for a new pesticide to determine the amount that is left in food after harvesting. You are asked to set up a quality management system, because your laboratory's management has decided that the quality of your group's analytical work should be assessed by an appropriate independent authority.

Which of the following Standards would you choose as an appropriate basis for your management system and why?

(i) ISO 9001:2000 certification;

(ii) ISO/IEC 17025:2005 accreditation;

(iii) GLP compliance.

9.2 Standards Available for Laboratories

The management standards available have already been introduced and their general features highlighted in Chapter 2. Some of the more specific components of the Standards will be covered in this chapter. For general chemical laboratories, the two Standards that are most frequently encountered are ISO/IEC 17025 and GLP and so these will be covered more fully in the following sections.

9.2.1 Good Laboratory Practice (GLP) Requirements

The background to this set of requirements has already been discussed in Chapter 2. It should be remembered that GLP relates to a study and not to specific tests. In fact, in some respects, it is very narrow in its scope of application as it is only a requirement for regulatory studies. The definition of a regulatory study was given in Chapter 2. A nominated monitoring authority will judge compliance with the GLP code of practice – in the UK, this is the Department of Health's GLP Monitoring Authority. A laboratory can announce that it is operating in accordance with the principles of GLP. However, regulatory authorities will require that the laboratory is included within the national GLP compliance programme and that the national GLP monitoring authority has inspected the laboratory, before they can accept the laboratory's study data. The principal requirements of the GLP scheme are summarized in the following sections.

9.2.1.1 Management

The responsibilities of personnel and the laboratory's management structure must be clearly defined, by means of organizational charts, job descriptions and 'curriculum vitae' for the personnel who are carrying out the study. There must also be up-to-date records of qualifications and of the training that the staff have received, including any records necessary to show their competence to carry out their work.

9.2.1.2 Role of the Study Director

A **Study Director** must be appointed, with overall responsibility for the study and for approving the study plan and any amendments to the study. The Study Director has the responsibility to oversee the technical aspects of the study and so must have appropriate qualifications and experience to be able to supervise the work carried out. The Study Director must ensure that the agreed protocols are followed and that any unavoidable deviations from the protocol are justified and fully documented. The Study Director is also responsible for the following:

- overseeing the recording of data from the study;
- ensuring that any computerized systems used in the study have been validated;
- the preparation of the final report and the archiving of all relevant material;
- indicating the extent to which the work complies with the study as contracted and the study complies with the requirements of GLP;
- signing the final report as an indication of acceptance of responsibility for the validity of the data.

There has to be a documented set of procedures for replacing a Study Director should that become necessary. In addition, for multi-site studies there has to be appointed a **Principal Investigator** for each site. Principal Investigators have to have the appropriate qualifications and experience to manage the delegated part of the study.

9.2.1.3 Quality Assurance Programme

There must be systems in place to monitor the study while it is in progress and to check that all systems are working in accordance with GLP requirements, to record any problems identified and to ensure that remedial action is taken. The person responsible for these quality assurance procedures must be independent of the study being audited. In large organizations, there may be a separate Quality Assurance unit but this will not be practical in smaller organizations.

9.2.1.4 Facilities

The facilities must be appropriate for the work being carried out. Procedures are required covering the receipt of test materials, their handling and storage, and how these substances are issued for use, so that the records of the use of the test material can be audited. The design of the test facility should provide separate areas for different activities to allow proper delivery of each regulatory study. Suitable accommodation is required for archiving the records and specimens generated by each study.

A test facility or an individual laboratory area within a test facility may be engaged in the conduct of both regulatory studies (GLP compliant) and other work (GLP not required). In such situations, measures must be taken to ensure that the GLP compliance of the regulatory studies is not compromised. The way to resolve this is for all of the work of the laboratory to be carried out to meet GLP requirements.

9.2.1.5 Equipment

Equipment must be suitable, maintained and, where appropriate, calibrated. Computer systems used to generate, store and retrieve data should be of appropriate design and capacity, validated and suitably located. If there are computerized systems used to control environmental factors, then these also require the same consideration. Issues relating to computerized systems are covered in Section 9.2.1.8.

9.2.1.6 Test Facility Working Procedures

Standard Operating Procedures (SOPs) should be properly authorized, documented and available to the staff carrying out the work. They must also be identified and controlled, so that all staff are aware of the current version and no outdated or altered copies can be used. SOPs should be regularly reviewed to ensure that they are still appropriate for the study programme. A system is required so that superseded versions of the operating instructions are filed and available, in order that studies carried out in the past using the SOPs in existence at the time of the study can be 'reconstructed' if necessary.

The status of test substances must be well-defined and recorded, including their identity, purity and properties. **Test items** and **reference items** should be stored separately to avoid contamination and reduce the potential for a 'mix-up'.

Reagents and solutions must be clearly identified, including shelf-lives and storage conditions if required. There must be adequate supplies for the length of the study.

9.2.1.7 Planning and Conducting a Study

Before a study begins, a study plan has to be prepared and approved. A study plan should make clear the title and purpose of the study, for whom the work is being carried out (and by whom), the timetable for the study and the test system that is to be used. The tests to be applied should be documented, as well as the statistical methods to be used to analyse the data. The system for storing the records from the study must be described, and accompanied by the names of those who are authorized to approve and issue the results of the study.

The study must be carried out in accordance with the study plan, and all data generated must be recorded promptly and signed or initialled and dated. Any corrections must identify who made the correction, when and why. The original data must remain visible. If any changes have to be made to the study plan, these have to be approved by the Study Director, in conjunction with the customer, and documented. The record has to explain why the change was made and the details of the change.

9.2.1.8 Computerized System Requirements and Electronic Records

Increasingly, computerized systems and electronic records are part of a laboratory's operations. However, records may be held in both paper and electronic format and the quality assurance required depends to some extent on the format of the definitive document. Because of space requirements involved with paper records and ease of retrieval of electronic records, the latter are gaining in popularity. The same requirements that we have for paper records, e.g. change control, readability and archiving, will still apply to electronic records. For this to be achieved, new procedures may have to be developed.

The computerized systems, both hardware and software, that form part of the GLP study should comply with the requirements of the principles of GLP. This relates to the development, validation, operation and maintenance of the system. Validation means that tests have been carried out to demonstrate that the system is fit for its intended purpose. Like any other validation, this will be the use of objective evidence to confirm that the pre-set requirements for the system have been met. There will be a number of different types of computer system, ranging from personal computers and programmable analytical instruments to a laboratory information management system (LIMS). The extent of validation depends on the impact the system has on product quality, safety and record integrity. A risk-based approach can be used to assess the extent of validation required, focusing effort on critical areas. A computerized analytical system in a QC laboratory requires full validation (equipment qualification) with clear boundaries set on its range of operation because this has a high impact. The risk imposed by using an inappropriate system is high. The word processing system used to generate documents and reports would require some validation to check it was correctly installed, together with robust change-control procedures. Inevitably, changes to the computerized system will be necessary at some time. To take this into account, there should be formal approval and documentation of any changes to the system during its lifetime. Computerized systems also fail from time to time so back-up facilities should be available and there should be documented procedures for dealing with system failures. If the change or a failure results in changes to the system, then some revalidation may be needed. The validation documentation relating to computerized systems has to be retained, along with the other documents relating to the study.

In terms of records in electronic format, the requirements are much the same as for paper records. There will be some extra work involved if the final record is in an electronic format. The integrity of records has to be maintained over the normal retention period. This can pose a problem for electronic records because not only has the integrity to be maintained but the readability also has to be assured. It may be necessary to move/transfer the records to a new/different system. Every five years, or as recommended by the media manufacturer, it is reasonable to check the stored tapes and transfer the information to a new tape. There should be checks built into the transfer process to ensure that the integrity of the material is retained. It is important that this process is validated to ensure that there is no loss of information or changes to the data.

Built into the design of computerized systems used to collect, process, report or store raw data electronically, there should be a means for tracking all events – an 'audit trail'. It is important that changes made to the data are visible without obscuring the original information and that there is a means of recording the reasons for the change. The identity of the person entering and/or changing the data should be associated with the event. This means there has to be a facility for timed entries and electronic signatures. Where there may be different time-zones involved in the study, then the time-zone associated with the entry should be unambiguous.

The security of data is essential for GLP studies, whatever the format of the documentation. The security of computerized systems in a laboratory environment is generally taken care of by the organization's security system. There may be extra precautions necessary if the equipment is located 'remotely'. Security of access is normally achieved by use of 'person-identification' and a password which changes on a regular basis.

The Study Director needs to have an understanding of the extent to which the computerized system impacts on the study results. The Quality Assurance team will also check that there are procedures in place to meet the GLP requirements for all stages of development, use and maintenance of the systems used. They will require 'read-only' access to the system.

Organizations that are regulated by the US Food and Drug Administration (FDA) have to comply with the Code of Federal Regulations 21 CFR Part 11 [5]. The FDA has published a guidance document on the scope and application of the regulations in relation to electronic records and electronic signatures [6]. The scope of Part 11 is restricted to records as required by **predicate rules** or the records that are required to demonstrate compliance with predicate rules. In this context, predicate rules are the underlying requirements set out in the Federal Food, Drug and Cosmetic Act, and the Public Health Service Act and FDA regulations (e.g. GLP), other than part 11.

All of the above may appear to involve a great deal of effort but the extent of the effort should be based on the level of risk involved. The starting point should therefore be a risk-based assessment, categorizing the systems as high, medium or low risk based on their impact on the quality of the final result. Electronic records that are generated by systems that are critical to the study should be examined in detail for the whole of their life cycle. The following process is a way to approach the risk-based investigation:

- determine the predicate rule requirements;
- identify the electronic records that require compliance;
- document current practice;
- identify where users have access to the data;
- list staff who can change the data;
- evaluate the impact the change has on the quality of results;
- use the information collected to define where an electronic audit trail is required.

This approach will limit the amount of work and identify where procedures need to be put in place to implement the changes that have to be made to the current practices, e.g. limit access, and the checks that need to be built into the process. All of this then should be documented as one or more SOPs. More detailed information is available on a website dealing with regulatory matters [7].

9.2.1.9 Final Report

The format of the final report from a study is closely defined and the Study Director is responsible for its production. The test item and name of the study must be clearly identified, along with the start and end date. It has to be accompanied by a statement from the Quality Assurance unit that the study and any critical aspects of the work have been conducted to the requirements of GLP principles. It will contain the names of all those involved with the study and their address; this includes the Study Director, the Principal Investigator, and all the scientists contributing to the report. The location of all of the materials stored (test samples and reference items) is archived.

9.2.1.10 Storage of Data

It is essential that all data generated and any other records or samples (if possible) are retained so that they are available for inspection at a later date. This means that there must be a secure and properly controlled archive, with an archivist to maintain it. The archive will contain a copy of the Study Plan, quality assurance reports, records of staff including their curriculum vitae (CV) and training records. Access to the archive has to be strictly controlled, and any additions to or withdrawals from the archive must be logged (see also electronic records in Section 9.2.1.8). Such removal or additions can only be undertaken by stated personnel.

9.2.1.11 Auditing

The UK GLP Monitoring Authority normally carries out inspections of laboratories every two years to assess whether they are operating to GLP principles. The initial inspection concentrates on ensuring that all of the systems required by GLP are implemented and is known as an 'implementation inspection'. Subsequent inspections confirm that all of the GLP principles are being applied. Any shortcomings identified during the inspection have to be remedied before the laboratory can be issued with a 'Statement of Compliance' which is the official document recording that the Monitoring Authority has agreed that the laboratory's quality system meets the requirements of GLP.

In addition to the regular biennial inspections, specific inspections of a particular study can be carried out at the specific request of regulatory authorities either in the UK or abroad. There may also be a surveillance visit to monitor the effectiveness of remedial actions arising from serious adverse inspection/findings.

9.2.1.12 Comparison with ISO/IEC 17025:2005

GLP compliance monitoring not only examines the procedures and practices used by the test facility to carry out regulatory work on chemical products (e.g. industrial chemicals, pharmaceuticals, veterinary drugs, food and feed additives, and cosmetics) but also evaluates performance. It is essential that the study can be reconstructed at some future date and that the integrity of the data generated can be demonstrated. In some respects, the requirements of GLP are more stringent than those of ISO/IEC 17025; these include recording and reporting of data. Management data have to be retained in an archive to allow complete reconstruction of a study. The other major difference is that GLP compliance is 'study-based', whereas ISO/IEC 17025 accreditation is 'test-based'. Some of the other differences are shown in Table 9.2.

Table 9.2 Some differences between GLP compliance and ISO/IEC 17025 accreditation

Requirements of GLP	Requirements of ISO/IEC 17025
Very specific responsibility of personnel	General statement for responsibilities of staff
Each study is audited by a Quality Assurance (QA) unit	Each test not subject to internal audit
Description of quality system is in SOPs	Description of quality system is in the Quality Manual
Each study signed off by the Study Director and the QA unit	Tests not signed off by QA unit but by person responsible for the work
Specific requirements for storage, retention and archiving of data and records	No specific requirements for the storage of reports and records
Inspection by legal authority required	Not a legal requirement
If there is a problem, then it has to be resolved through a court of law	There have to be documented complaints procedures
Storage of samples according to local regulatory requirement	Storage of samples and data according to laboratory policy or until the customer accepts the results

9.2.2 ISO/IEC 17025 Requirements

The first edition of ISO/IEC 17025, 'General Requirements for the Competence of Calibration and Testing Laboratories', was produced at the end of 1999. The second edition, ISO/IEC 17025:2005, was produced to bring it in line with ISO 9001:2000. Before 2000, many countries operated their own accreditation standards. The introduction of ISO/IEC 17025 means that now there is one internationally accepted document for national accreditation bodies across the world on which to base their standards for laboratory competence. There are two main sections in the Standard, Section 4 dealing with the management requirements and Section 5 dealing with the technical requirements. The management requirements are very similar to those in ISO 9001:2000. Compliance with the requirements of ISO/IEC 17025 indicates that a laboratory is operating a management system that meets the principles of ISO 9001. The laboratory does not have to be certified to ISO 9001:2000. The requirements of ISO 9001:2000 will be dealt with in more detail in Section 9.2.3. There are fifteen main headings in Section 4 of the ISO/IEC 17025 Standard, and Annex A of the Standard has a table cross-referencing the clauses to ISO 9001:2000. In the order that they appear in the Standard, they are: management requirements; management system; document control; review of requests and tenders; subcontracting of tests and calibrations; purchasing services and supplies; service to the customer; complaints; control of nonconforming tests and/or calibration work; improvement; corrective action; preventive action; control of records; internal audits and management review.

A laboratory must have a defined quality management system, which is described in the Quality Manual. The procedures for auditing and reviewing quality management must be documented and put into practice so that the laboratory can demonstrate that standards of quality are being maintained, monitored and are still appropriate. Note that these are internal matters (see Sections 9.4 and 9.5).

In addition, a system for making sure staff are appropriately qualified and trained for the work that they are doing must be in place. This will enable an auditor to see clearly the demonstrated competence of the staff and how this has been checked. The requirements for all major items of equipment must be listed, to ensure that the equipment in use is suitable for the task, is in working condition and, where necessary, is calibrated. For all of the instrumentation there needs to be a documented schedule for maintenance. Measurements must be traceable, that is, the laboratory must be able to show how the calibration of measurement instruments is traceable to National or International Standards. Where this presents practical problems, as in some chemical measurements for example, interlaboratory comparison and the use of reference materials (and preferably Certified Reference Materials) will be required.

Methods and procedures, including sampling, sample handling, analysis and the estimated uncertainty of the final result, must be appropriate for the work carried out. All of the methods used, standard and non-standard, must be fully validated and documented. The extent of validation has to be considered on a case-by-case basis. The integrity of all analytical data must be protected at all times so that raw data can be inspected at a later date if required.

The laboratory accommodation and environment must be suitable for the analyses being carried out. For example, laboratories carrying out analysis for trace levels of metals must be able to demonstrate that there is no risk of contamination from the specific metals in the vicinity of where the analyses are being carried out or where the samples are stored.

Test samples must be uniquely identified and prevented from deteriorating before the analysis is performed. Procedures to authorize ultimate disposal of samples must also be documented.

A detailed and comprehensive system of record keeping is necessary, including, for example, worksheets, notebooks, computer output and reports, and all of these should be retained for a reasonable period of time or as required by the customer. A period of six years is often chosen. The content of reports and certificates is tightly defined, to ensure that customers receive all relevant information and that the laboratory does not make exaggerated claims about which parts of its work have been accredited. A documented system for dealing with any customer complaints and for informing customers if discrepancies in results are subsequently discovered must be available and in place.

Finally, the laboratory's freedom to subcontract tests or make use of outside services is strictly defined, to ensure that work placed with an accredited laboratory is not 'farmed-out' to a laboratory with inadequate quality procedures.

In summary, the Standard requires that a laboratory must clearly document its procedures, ensure that these are carried out correctly and be able to demonstrate to a third-party that they are under control and have been carried out correctly.

9.2.2.1 Management Requirements

The Standard does not require all organizations to implement a quality management system that is identical with all others. The quality management system has to be appropriate to the scope of the organization's activities, i.e. 'fit for purpose'. There are a few specific requirements. These include the appointment of a person who has the responsibility and authority to ensure that the quality management system is followed. This person is usually given the title of Quality Manager and has direct access to the most senior manager within the company, e.g. the Chief Executive Officer. The Standard specifies that the testing or calibration activities of the laboratory have to be carried out in such a way that not only satisfies the requirements of the Standard but also of the customer or regulatory body. Large organizations will be involved in a variety of activities other than testing and/or calibration. The organization chart has to show the responsibilities of staff who are involved with or can influence the laboratory activities. This identifies potential conflicts of interest so that they can be prevented from happening. There also have to be in place policies to ensure that the confidentiality of customers' information is retained.

9.2.2.2 Technical Requirements

Many of the technical requirements of the Standard are covered in Chapters 4 to 7. The analytical requirements, including choosing a method and method validation, are covered in Chapter 4. The other measurement requirements, such as calibration, traceability and equipment qualification, are dealt with in Chapter 5. Some of the general issues not covered elsewhere are mentioned in the following sections. It has already been mentioned that staff should be trained and proven to be competent to carry out the testing. This applies to permanent and contracted staff. The laboratory should have a 'job description' for all members of staff. There are more stringent requirements on staff who are also able to provide customers with opinions or interpretation of the results.

9.2.3 ISO 9001 Requirements

As has already been mentioned in Chapter 2, ISO 9001, 'Management Systems – Requirements', is increasingly being adopted by laboratories to cover the aspects of their business that are not laboratory based. This is because this Standard is more about controlling the process and service enhancement rather than technical issues. It requires continuous improvement, demonstrating that quality is not a static process. The requirements for such matters as documentation, document control, purchasing and management responsibilities are much the same as for ISO/IEC 17025. This means that additional effort is unlikely to be required. There are no technical requirements specified other than that the service provided must be 'fit for purpose'. Certification does not give any assurance of the competence of the laboratory staff.

9.2.3.1 Quality Policy

The top management of the organization produces the quality policy. This sets out the overall objectives of the quality management system. It should include reference to the management's commitment to good professional practice, compliance with the quality standard, and a framework for establishing and reviewing quality objectives.

9.3 Quality Manual and other Documentation

In terms of documentation, the requirements of ISO/IEC 17025, ISO 9001 and ISO 15189 are very similar and so will be dealt with together. For a quality management system to be effective, all of the components (policies, systems, programmes, procedures, instructions, etc.) must be clearly documented so that everyone in the organization knows what is expected of them. Figure 9.1 shows

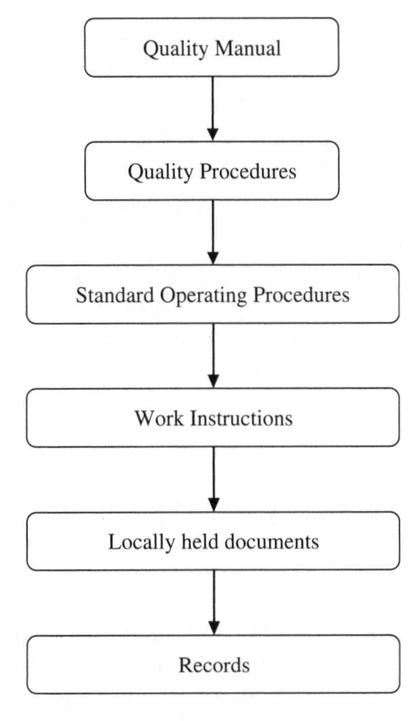

Figure 9.1 Hierarchy of quality documentation.

a flow diagram of the different types of documentation required in an analytical laboratory.

The main component of the documentation is usually referred to as the Quality Manual although it may have another name. The exact content and format of the manual is not specified in either ISO/IEC 17025 or ISO 9001. There is a suggested table of contents of a Quality Manual given in ISO 15189. An example of the content of a Quality Manual is shown in Table 9.1. The Quality Manual sets out the structure of the management system and the management's policy on key aspects of the system. For ISO/IEC 17025 it must clearly set out the laboratory's scope of accreditation, i.e. the range of tests for which the laboratory has been accredited. It must also include management and technical responsibilities, and the laboratory's operating and quality control procedures. The Quality Manual will contain training instructions but training records will be kept locally, e.g. in the section where the person works. Chapter 8 deals with documentation in more detail.

9.4 Audit

A system for auditing and reviewing quality procedures is a specific requirement of ISO/IEC 17025, ISO 9001 and GLP. This is a critically important aspect of any quality management system so we will consider these activities in some detail. The first, and most important, thing to realize about Quality Audit and Management Review is that they are two completely different activities. This section deals with auditing while Section 9.5 deals with management review.

Quality Audit is a process, carried out at regular intervals, of testing the management system in use in the laboratory to check if the processes are compliant with the Standard, effective, documented and being adhered to by the working staff. A quality audit should check that you have been carrying out your job as set down in the laboratory's written procedures. Quality Audit is the responsibility of the Quality Manager who will plan, and undertake audits and select other specialist auditors. There are different types of audit, including internal audits, second-party audits and third-party audits.

DQ 9.3

What do you think is meant by these different types of audit?

Answer

You may have already experienced an audit so would be able to identify an *internal audit*. This is an inspection carried out by staff of the laboratory but who are independent of the work activity carried out in the area being audited, i.e. the auditee. Both the *second-party* and *third-party audits* are *external audits* [8]. *Third-party audits* are carried out by independent auditing organizations, such as those providing certification/registration/compliance/accreditation of conformity to a particular Standard such as ISO/IEC 17025. *Second-party audits* may be a less familiar term. This is an audit carried out by a person or persons external to the organization being audited but who has an interest in the organization. This may be a customer or a representative of a customer.

It may seem unnecessary to have audits especially when there seem to be so many different types of audit. However, in spite of all of the documented procedures there are problems that can arise in analytical laboratories due to changes in staff, procedures, equipment, sample type and number of samples. For a laboratory to provide a consistent standard of quality in the face of all of these regular operational changes, Quality Audits need to be carried out. Audits will identify the problems which are expected to emerge, and provide a system to put them right.

For laboratories seeking external accreditation of their work, audits must be planned and written down so that the laboratory can show that they are being carried out. You should note that a laboratory will not be criticized for finding problems when they audit their work. The important thing is to be on the lookout for problems, to find them where they exist and to put them right. No laboratory should rely solely on periodic assessments by an external body to ensure that standards of quality are continuously maintained. There are a number of reasons for this.

Third-party assessments are sampling exercises, carried out by assessors who may be unfamiliar with the detailed operational procedures they are considering. Assessors try their best, but they cannot guarantee to uncover all the problems that a laboratory may have. External assessors' findings should be regarded as indicative of the types of activities that need to be re-examined by the laboratory, rather than as merely specific instances of non-conformity with a Standard. External assessments are rather like Ministry of Transport (MOT) certificates obtained for motor vehicles – they are correct on the day that they are issued, that is, they give a 'snapshot' view on a particular day. The vehicle may still break down the following day! If third-party assessments of laboratories are conducted annually or biennially, there is ample time for things to go wrong between the visits.

In large laboratories, where dozens or even hundreds of different types of analyses are carried out, an additional internal process to test the quality system is particularly important. An external assessment team can only hope to observe and assess in detail a relatively small number of the analyses during each visit, so there may be years between any one of a laboratory's externally assessed analyses being re-examined in detail. Laboratories may not have entered all of the analyses that they carry out in their scope of external accreditation. If they wish to ensure quality is maintained in these other areas of their work, an internal

Quality Audit is the only option available to them. It is therefore vital that, to ensure standards of quality are maintained, a laboratory must operate its own internal audits to test its quality management system. If the internal system of audits is not operating adequately, external assessment visits are liable to be traumatic experiences, awaited with trepidation and producing unexpected and embarrassing nonconformities in several areas of operation.

However, if the internal system of audits is operating satisfactorily, the laboratory's management can await external assessments confident in their systems, which they themselves have already tested. The external assessment process can then become more like a consultancy, with informed discussion between external auditors and laboratory staff over current 'best-practice' for maintaining and improving quality. So, although external assessments offer a valuable insight into a laboratory's procedures, and an independent recognition of the quality of operations, they cannot by themselves ensure continuing quality within the laboratory. For the laboratory to maintain and improve its quality of operation, it must continually test and re-examine its own management system. A systematic and regular process of internal quality audit offers a structured route to achieve this.

9.4.1 Responsibility for Internal Quality Audits

As was mentioned earlier, auditing is the procedure used to test whether a laboratory's management system is working as intended. Internal quality audit is the responsibility of the Quality Manager, who must have direct access to senior management to report his/her findings and recommendations. As some of the recommendations from the audit may be difficult or costly to implement, the Quality Manager must also be of sufficient standing within the organization to ensure that any actions necessary to protect the quality standard of the laboratory are carried out.

In a small laboratory, the Quality Manager may personally be able to carry out the Quality Audit. In a larger laboratory, however, it will probably be more appropriate to share the task of auditing with a number of auditing officers, who report their findings to the Quality Manager. All auditing officers however, must be independent of the activities which they are asked to audit. If it is impractical to use internal auditors, a laboratory can opt to employ external auditors to carry out their internal Quality Audit. Finally, it is the responsibility of the Quality Manager to ensure that any nonconformity identified during the audit is satisfactorily dealt with and to forward the results of the Quality Audit to the laboratory's management team for consideration as part of the Quality System Review. A nonconformity is any non-fulfilment of a need or expectation that is stated or implied in the documentation relating to the Standard being audited. Examples of some nonconformities are given in Section 9.4.7.

9.4.2 Planning of Internal Quality Audits

Each area of operation should be audited periodically, usually once every twelve months. However, new areas may be audited on a more frequent basis. Ideally, they are planned 'mid-way' between external audits. This process should be planned well in advance and written down in the Quality Manual as a structured audit programme, covering both the timing and the coverage of the audit process. The audit programme can be of two types, either:

(i) a 'rolling programme', organized so that different activities within the laboratory are audited each month in a series of visits, so, for example, in January training is audited in all parts of the laboratory, in February calibration, in March equipment, etc.

or:

(ii) a complete audit, covering all activities of the laboratory in one visit, so that, for example, in January and July all aspects of the laboratory's work are audited in the course of one or two days.

A laboratory is free to decide which scheme is most appropriate for its own circumstances. Laboratories entering the audit process for the first time may prefer to carry out an initial complete audit to establish if problems exist, followed, after an appropriate interval, by a rolling programme to maintain standards. When auditing to ISO/IEC 17025, the auditor should plan to witness at least one method being carried out.

9.4.3 Training of Auditors

Unless there are some members of the staff of the laboratory who have been trained as assessors by one of the accreditation bodies and who are therefore used to performing quality audits, the auditors will need to be instructed by the Quality Manager as to how they should go about their task. Auditors will need to be instructed not only on which aspects of the work should be examined with reference to the requirements of the management system and the relevant Standard, but also on how the audit should be approached. There needs to be a real test of the systems being audited, but without degeneration into a 'nit-picking' or 'point-scoring' exercise that will lose the goodwill of the staff being audited. The objective of the audit process should be to improve the level of quality in the operation of the laboratory, so open discussion and constructive suggestions should be the rule, rather than negative criticism. The personality and presentational skills of the staff selected to be auditors will therefore be extremely important, as well as their technical knowledge. The Standard ISO/IEC 17025 requires it to be a person with the demonstrated personal attributes and competence to conduct the particular audit, who will act as a guide as well as a 'policeman'.

SAQ 9.2

Consider your colleagues in your own laboratory – which of them do you think would be suitable to act as an auditor, and who would be unsuitable? Consider the characteristics you would look for in a potential auditor.

9.4.4 Conduct of Internal Quality Audits

It is important that before the audit is started, everyone involved is clear about what is going to happen. This will include the structure of the audit, how matters relating to nonconforming work will be documented and how the findings will be reported. The timescale for any remedial actions should be explained as this forms a critical part of the feedback and corrective action process. During the audit, a member of staff familiar with the work being audited should accompany the auditor. This person is present to explain the laboratory procedures to the auditor and, where the auditor wishes to make an observation, agree the factual basis of the observation on behalf of the head of the section involved. As a general rule, the audit against the Standard ISO/IEC 17025 can be split into two parts – management (process and administrative) and technical (competency, methods, traceability and uncertainty). The most important aspect of auditing is to be able to follow the process from sample receipt to report delivery, without encountering 'leaps of faith'.

It must be stressed that, to be effective, internal auditors should not make allowances for any operational problems within the laboratory, which they themselves may also be victims of, such as cramped accommodation or inadequate fume cupboards, for example. If these are factors that could affect the quality of the laboratory's results, they must be recorded as a problem as part of the audit. The auditors must therefore prepare a list of all their findings, good and bad, for the Quality Manager. This approach may force laboratory managers to go back to some long-standing and difficult problems and reconsider them as issues of quality. This can often prove to be the stimulus to get long-awaited improvements carried out. There are therefore occasions when working scientists in a laboratory welcome the visits of auditors as an opportunity to explain why money needs to be spent on improving their equipment or facilities. At the end of the audit, nonconformities should be documented and the appropriate corrective action agreed with the head of the section concerned or their authorized deputy. It is essential that the timescale for the corrective action is agreed and recorded. At the agreed time, the auditor will then return and check whether the corrective action has been completed. If the auditor finds that the corrective action has not been carried out, the person responsible will be clearly identified and so they will have to be ready with some plausible explanations! The auditors' reports should then be passed to the Quality Manager to note and compile into the report to go forward to the Quality System Review meeting. The laboratory staff are expected to investigate ways of preventing such noncompliances happening in the future. The auditing process provides an opportunity for staff to identify improvements to the process or system.

Examples of Quality Audit report forms are shown in Tables 9.5 and 9.6 at the end of this chapter. Table 9.5 shows Report Form 1 which includes a record of what has been examined, so that subsequent audits can examine other aspects of the laboratory's operations. Note that when improvement actions are required, the form records not only what needs to be done, but also by when it must be completed, as shown in Table 9.6 (Report Form 2). The form is signed by the auditor and the responsible person from the area being audited. In the UK, the accreditation body UKAS refers to the corrective action as 'improvement action'. This further emphasizes the continual improvement aspect of the ISO/IEC 17025 and ISO 9001 Standards. The examples shown in Tables 9.5 and 9.6 will probably have to be modified to meet a laboratory's particular requirements. It should also be remembered that both report forms should indicate (usually as a footer or header) the title of the document, its issue date, issue number, who authorized the document, page number and total number of pages.

9.4.5 Coverage of Internal Quality Audits

DQ 9.4

What do you think should be looked at in the course of a Quality Audit?

Answer

All aspects of the laboratory's work which might affect the validity of the final result should be inspected. This will include, for example, documentation, equipment, calibrations, methods, materials, record keeping, sample recording, labelling, quality control checks and log of daily checks, among many others. Some aspects, however, are outside the scope of such an audit, such as safety and security matters, which usually have separate arrangements for auditing.

To ensure that all aspects of the laboratory's activities are covered over a set period of time, a spreadsheet of activities can be prepared. This shows when each area is due for audit. In order to have a consistent approach, a check-list approach may be adopted. An example of a check-list of aspects which should be examined as part of an internal quality audit is shown in Table 9.5. This can be used as the basis for an audit against any of the Standards already mentioned in this chapter. Not all of the parts will be appropriate for all of the Standards. Table 9.7 expands on what is included in Table 9.5 and can be used to remind auditors and laboratory staff of the things that can affect the quality of a result.

To enable the auditors to carry out their function effectively, it is important that they should understand the basis of the testing being carried out. The auditors should therefore have access to the relevant documentation in advance of the audit visit, to enable them to become familiar with the principles and practical details of the analyses to be audited. It is not unknown for an assessor or auditor to discover that, although the documented test procedures are being carried out exactly as specified, they are not in fact appropriate for the samples being examined. For example, a method developed and validated for water analysis, where almost 100% recovery of analyte can be achieved, might then have been applied to samples of sludge, where recoveries are perhaps only around 10%, without anyone having checked that the analytes can actually be recovered from this new substrate. This emphasizes the need to ensure fitness for purpose and particularly to validate methods for particular types of sample. Similarly, it is sometimes found that small variations in the sampling procedure carried out before the analysis is started are rendering the analytical results invalid. A full understanding of the basis of the analytical approach is therefore necessary in order to allow the auditor to recognize such inconsistencies. Many of these factors have already been discussed in detail in previous chapters.

9.4.6 The 'Vertical Audit'

In addition to the detailed checking of procedures, as set out in the check-list, 'audit trails' are particularly valuable. This is a 'vertical audit' and refers to the examination, in chronological order, of all records relating to a particular sample, from the moment of receipt through the various analyses carried out, to the reporting of results and the ultimate disposal of the sample. These vertical audits are therefore also sometimes referred to as 'birth-to-death' audits. In contrast, a 'horizontal audit' covers some specific activity across a number of analyses – this may be, e.g. solid-phase extraction or HPLC.

This type of audit of a sample's history, asking for all records, charts, spectra, calculations, etc. to be produced can often bring to light problems which the 'horizontal audit' of particular activities will not reveal. It may, for example, reveal the use of equipment which has not been listed as part of the method, and which may therefore not have been included in the laboratory's calibration procedures. During such an audit, the auditor should re-interpret all of the raw data and redo any calculations that have been carried out, so as to be sure that the reported results are correct. In addition to helping to find any errors, examining the raw data will often highlight any spurious claims as to the detection limits and uncertainties of the measurements being made.

9.4.7 Types of Nonconforming Work

During an audit, a failure to comply with any component of the relevant Standard(s) is recorded as *nonconforming* work or observations. There are a number of categories of nonconforming work, which depend on the nature of the failure. There is no agreed way of categorizing nonconformities, so here only descriptive terms are used. The extremes of the scale are, at one end, minor failures which present a low risk to the quality of the work and at the other end major failures that would put the quality of the test at risk. A minor category of the nonconforming work may be allocated for a minor failure to comply with the requirements of the Standard. It may also be recorded for repeated incidences of a minor failure to comply with the Standard, provided that there is no associated quality risk. Examples include errors in data recorded in workbooks – corrected but not initialled, an organization chart not up-to-date, no calibration label on an item of equipment, a reference standard not calibrated by the due date but no calibrations had been performed based on this item since the recalibration date, etc.

The major category of nonconforming work is allocated for any failure of a system to comply with the requirements of the Standard which could lead to invalidity of test results. Examples include absence/non-implementation of a document control system, absence/non-implementation of a procedure for internal audit or management review, staff not technically competent to perform particular tests and failure to control the quality of test data.

The intermediate category of nonconforming work is allocated when a number of related minor failures are observed, which together are judged to be an unacceptable quality risk, without constituting an overall system failure in the areas concerned.

The consequences of nonconformities are different for the different categories. Minor nonconformities will be noted and checked at the next assessment but will not normally appear in the written report. Major nonconformities can result in a total suspension of accreditation or suspension of accreditation for particular work. The intentional misuse of the accreditation body logo or mark is considered to be a serious nonconformity. If, after a previous warning, a laboratory continues to issue test reports showing the accreditation logo when the tests are outside the scope of the accreditation it can result in withdrawal of accreditation. Accreditation can be withdrawn for a particular area if it is found that accommodation is inappropriate and it is impossible to avoid 'cross contamination' of samples.

The nonconforming work that will require evidence of implementation of corrective and preventive action within a given timescale will include matters, such as, no corrective action taken when the results from a round of a Proficiency Testing scheme indicated the laboratory's result was an outlier, or the competency records of staff do not indicate they are competent to do the accredited work. Listings of nonconformities can be found in a publication produced by the International Laboratory Accreditation Cooperation (ILAC) [9].

A GLP audit report will identify any deviations from the GLP principles or other deficiencies found at the time of audit. These deficiencies may be 'minor' (do not affect the validity of the study being carried out by an organization) or 'major' (may affect the validity or integrity of the study). Examples of deficiencies include the calibration procedure for a HPLC system is inadequate in that it does not include integrator and detector linearity tests, injector reproducibility and accuracy of temperature settings for column heater and detector; standard weights, sample weights and calculations are not recorded.

9.5 Management Review

Internal Audit and Management Review are complementary activities but are different from each other. Internal management review is the responsibility of the laboratory management team, supported by the Quality Manager. At least once a year, top management must meet and consider an annual report produced by the Quality Manager. At this meeting, all aspects of quality, from top-level objectives and risk control to management standards, resources, level and scope of work and whether quality is at a satisfactory level, are reviewed. The Quality Manager's annual report is a very important part of the annual quality management system review. The Quality Manager then has the responsibility of recording the outcome of the review, including recommended actions, and ensuring that these actions are put into effect within the agreed timescale. Table 9.3 indicates who is expected to attend the management review meeting.

Table 9.3 Attendance list at a Management Review meeting

Chief Executive/Managing Director[a]
Finance Director[b]
Technical Director
Technical Manager
Senior managers
Quality Manager
Meeting Secretary[c]

[a] The Chief Executive/Managing Director is present to ensure that authority for any actions carries the highest authority.
[b] The Financial Director is present to ensure that any financial implications of actions can be discussed.
[c] The Meeting Secretary is present to ensure that the proceedings of the meeting are recorded.

9.5.1 Organization and Coverage of Management Review

The Management Standards followed by laboratories all require that the laboratory's quality management system is reviewed periodically to ensure that it is still suitable and effective and to introduce necessary changes or improvements. The usual frequency for quality Management Review is once every twelve months.

DQ 9.5

Suggest what pieces of information about quality in the laboratory should be available for the laboratory's management to consider at their annual Management Review meeting as evidence of the standard of quality.

Answer

The management team should examine all relevant information that is available to them. The exact list will vary between Standards and will usually be agreed between the laboratory and the accreditation/certification body. You will probably be able to think of a number of things to include. Table 9.4 shows a list that would form the basis of an agenda for such a meeting.

Table 9.4 Inputs to a Management Review

(1)	Actions from previous year's review
(2)	Changes in volume and type of work
(3)	Changes in International Standard requirements
(4)	Changes to Quality Manual
(5)	Suitability of policies and procedures
(6)	Results from internal audits
(7)	Results from external audits (second- and third-party)
(8)	Review of performance in laboratory intercomparisons or Proficiency Testing schemes
(9)	Status of preventive and corrective actions
(10)	Review of customer feedback, complaints and compliments
(11)	Training
(12)	Review of cost of quality
(13)	Recommendations for improvement

The top management team will use this information to conduct a review of current procedures to ensure that they continue to be satisfactory. The resource implications of quality decisions will need to be discussed at this meeting. Quality costs money and the Management Review is where financial matters can be discussed, including how much money has been allocated for maintaining or improving quality. If a laboratory's management sets store by their commitment to quality, they must be prepared to accept and approve the financial implications that result. This is the reason for the choice of attendees at a Management Review meeting listed in Table 9.3.

9.6 Responsibilities of Laboratory Staff for Quality

Responsibility for maintaining, operating and improving the laboratory's quality management system lies with every member of the laboratory's staff. A laboratory's quality management system can only really be successful if everyone is playing his or her part in the system. The next sections show how the different groups within the laboratory can contribute to the overall effectiveness of the quality management system.

9.6.1 Laboratory Management's Responsibilities for Quality

The management of a laboratory has the initial responsibility of deciding on the laboratory's quality policy and selecting the appropriate Standard (or Standards) for their laboratory to adopt. They must then make available the resources that will be necessary to put the requirements of the Standard(s) into practice, including appointing an appropriate person to be the laboratory's Quality Manager.

The laboratory's quality management system will then be drawn up in the form of a Quality Manual, and the management will be required to approve this manual as the written detail of how their quality policy is put into practice. Management then has a continuing responsibility periodically to re-examine the laboratory's quality management system to see if it is still appropriate to the needs of the laboratory's work programme. This is usually carried out by means of the Management Review meeting, although there are likely to be a series of particular quality-related items brought to the attention of the management during the course of each year. Management's final responsibility is to supply the resources necessary to maintain the quality management system at the required level.

9.6.2 The Quality Manager's Responsibilities

The Quality Manager acts as the focal point for quality issues within the laboratory. The Quality Manager is therefore responsible for ensuring that the laboratory is familiar with the requirements of the relevant Standard(s). This person is also responsible for drawing up and maintaining the Quality Manual, which sets out how the laboratory's quality management system is operated in practice. The Quality Manager has to organize the laboratory's system of audits of the management system and to ensure that any problems identified by the audits are corrected within an agreed timescale. The Quality Manager then prepares all of the relevant material for consideration at the Management Review meeting and ensures that the decisions reached at this meeting are carried out. In a laboratory that is accredited by an independent accreditation body, the Quality Manager will also be responsible for liaising with the accreditation body and for making the necessary arrangements for their periodic assessment visits to the laboratory.

9.6.3 Responsibilities of Individual Members of Staff

All members of staff of a laboratory are responsible for ensuring that they are familiar with the quality management system, as set out in the Quality Manual and any supporting documentation. They are expected to follow the procedures set out in the Quality Manual. However, this does not mean that they should merely become 'robots', with no freedom of choice or expression in their work. They should instead be using their practical expertise and experience to suggest improvements that could be made to the laboratory's systems to reflect changes in customers' requirements, improvements in technical equipment and all the other changes, which continually occur in analytical work. It should always be borne in mind that the Standards are not intended to prevent change, but they do require that changes are handled in a structured way. Change is a constant requirement of any dynamic system but, if introduced in a haphazard manner, can cause confusion and error. Any quality management system has therefore to be able to accommodate changes which will improve the way the laboratory operates but must ensure that the changes are considered, approved, documented and introduced in a controlled manner.

Summary

This chapter describes how a laboratory manages the quality of its work. In Chapter 2, an indication was given of the Standards a laboratory might select. This chapter compares and contrasts such Standards and sets out how a laboratory chooses the most appropriate Standard to demonstrate the quality of their work. Many components make up a management system and each one of them is described, including the purpose and conduct of audits, internal and external. Examples are given of some of the documentation required.

References

1. 'Quality Management Systems – Requirements', ISO 9001:2000, International Organization for Standardization (ISO), Geneva, Switzerland, 2000.
2. 'Good Laboratory Practice', Statutory Instrument 1999, No. 3106, Her Majesty's Stationery Office (HMSO), London, UK, 1999.
3. 'General Requirements for the Competence of Testing and Calibration Laboratories', ISO/IEC 17025:2005, International Organization for Standardization (ISO)/International Electrotechnical Commission (IEC), Geneva, Switzerland, 2005.
4. 'Medical Laboratories – Particular Requirements for Quality and Competence', ISO 15189:2007, International Organization for Standardization (ISO), Geneva, Switzerland, 2007.
5. 'Code of Federal Regulations, Title 21, Food and Drugs, Part 11 Electronic Records; Electronic Signatures; Final Rule', *Federal Register*, **62**(54), 13429–13466, 1997.
6. 'FDA Guidance for Industry, Part 11, Electronic Records; Electronic Signatures – Scope and Application' (Final version, August 2003), Center for Drug Evaluation and Research (CDER), US Food and Drug Administration (FDA), Beltsville, MD, USA, 2003.
7. Labcompliance, Oberkirch, Germany [http://www.labcompliance.com].
8. 'Quality Management Systems – Fundamentals and Vocabulary', ISO 9000:2005, International Organization for Standardization (ISO), Geneva, Switzerland, 2005.
9. 'Guidelines on Grading of Non-conformities', ILAC G20:2002, International Laboratory Accreditation Cooperation (ILAC), Silverwater, Australia, 2002.

Table 9.5 Quality audit report form 1

Report Form 1

Instructions for use

1. All audits should be conducted to ensure coverage of and compliance with the appropriate Standard, e.g. ISO/IEC 17025:2005 and ISO 9001:2000.
2. The report form should be completed by the auditor and passed to the Quality Manager immediately after the audit.
3. Part A is applicable to all activities but Part B is applicable to testing and calibration areas only.
4. In both Parts A and B, specify which items or documents have been checked. If the item is not applicable, this should be indicated; the space should not be left unmarked.
5. Observations leading to corrective or preventive actions should be recorded on Report Form 2 (see Table 9.6).

ACTIVITY

TEAM REPRESENTATIVE

AUDITOR

Report Form 1

Part A

To be completed for all activities

	Controlled documents held *All documents should be current versions*
Quality Manual and Quality Procedures (QPs) *Note which QPs from the Quality Manual are applicable*	
Work Instructions *Note applicability, document control*	
Local Organization *Note organization, responsibilities*	
Training Program *For all staff, 'fitness for purpose', proof of competence*	
Other *Any other local internal documents*	
External Documents *e.g. Regulations, note document and version control*	
Contract Review *Note samples/files reviewed, records of discussions with customers*	
Customer Feedback	
Customer Reports	
Local Annual Management Review	

	Procedures audited *Specify which QPs, WIs, SOPs examined*
Methods or Procedures *Specify reference*	1. 2. 3.
Vertical Audits *Note which files/samples*	1. 2.

	Records *In all instances, give references of records audited*
Files	1. 2. 3.
Computerized Systems, IT, Databases, etc. *Note version control and change control*	
Training Records *State name or reference* *Check staff have appropriate qualifications, experience and training to carry out the work they do*	1. 2. 3.
Complaints *Evidence of resolution of complaint, placed entry on complaints database*	
Other	

Part B

To be completed for testing and calibration work only

	Equipment and calibration *Note references and relevant details*
Equipment List *Availability*	
Manuals	
Log books	
Labelling and Calibration *Check calibration certificate, indicate equipment identification details*	
– balances	
– thermometers	
– other equipment	

	Methodology and quality control *Note methods audited*
Methods/SOPs	
Method or Procedure observed	
Validation Data	
Calibration Data	
Use of Control Charts	
Intercomparison or PT Results	
Labelling (e.g. expiry dates)	

	Other QC
Uncertainty of Measurement	
Traceability of Results	
Other	

	Data handling (Registers or LIMS)
Sample Registration	
Raw Data, Worksheets, etc.	
Reports *Traceability of data and transcriptions errors*	
Other	

	Cause, corrective action and preventive action *Note (as an observation) recommendations*

	Housekeeping *Note general impressions*

Part C

To be completed for all activities

The auditor should summarize the findings, both good and bad, noting the number of observations made on Audit Report Form 2 (Table 9.6).

Auditor's signature ______________________________ **Date** ____________________

Auditor's name ______________________________

Table 9.6 Example of an audit report form where improvement/corrective action is required

Audit Report Form 2

Work Area ______________________ **Auditor** ______________________

Date of Audit ______________________ **Laboratory Representative** ______________________

This form should be completed by the auditor and copied to the Quality Manager, Division Head, and laboratory representative. At the expiry of the agreed timescale, the auditor should check whether improvement/corrective action has been completed and a report copied to the Quality Manager.

Observation	**Agreed improvement action and timescale**	**Improvement action checked**

Date ____________ Signature of Auditor ______________________ Signature of Laboratory Representative ______________________

Name of Auditor ______________________ *Name of Laboratory Representative* ______________________

Footnote with file name, page number and total number of pages

Table 9.7 Checklist for quality audit

Indication of areas of particular importance to a chemistry laboratory and which should be considered before and during an audit. This is not a definitive list and is not complete for all circumstances

1. Staff

- Staff have the appropriate blend of background, academic or vocational qualifications, experience and on-the-job training for the work that they do.
- On-the-job training is carried out against established criteria, which wherever possible are objective. Up-to-date records of the training are maintained.
- Analytical procedures are only carried out by authorized analysts.
- The performance of staff carrying out analyses is observed by the auditor.

2. Environment

- The laboratory environment is suitable for the work carried out.
- The laboratory services and facilities are adequate for the work carried out.
- There is adequate separation of high-level and low-level work.
- The laboratory areas are sufficiently clean and tidy to ensure that the quality of the work carried out is not compromized.
- There is adequate separation of sample reception, preparation, clean-up, and measurement areas to ensure that the quality of the work carried out is not compromized.

3. Equipment

- The equipment in use is suited to its purpose.
- Major instruments are correctly maintained and records of this maintenance are kept.
- Appropriate instructions for use of equipment are available.
- Traceable equipment, e.g. balances, thermometers, glassware, timepieces, pipettes, etc. are appropriately calibrated, and the corresponding certificates or other records demonstrating metrological traceability to National Standards are available.
- Calibrated equipment is appropriately labelled or otherwise identified to ensure that it is not confused with uncalibrated equipment and to ensure that its calibration status is clear to the user.
- Instrument calibration procedures and performance checks are documented and available to users.

Table 9.7 (*continued*)

- Instrument performance checks and calibration procedures are carried out at appropriate intervals and show that calibration is maintained and day-to-day performance is acceptable. Appropriate improvement/corrective action is taken where necessary.
- Records of calibration, performance checks and improvement/corrective action are maintained.

4. Methods and Procedures

- In-house methods are fully documented, appropriately validated and authorized for use.
- Alterations to methods are appropriately authorized.
- Copies of published and official methods are available.
- The most up-to-date version of the method is available to the analyst.
- Analysts are (observed to be) following the methods specified.
- Methods have an appropriate level of advice on calibration and quality control.

5. Chemical and Physical Standards, Calibrants, Certified Reference Materials and Reagents

- The reference materials required for the tests are readily available.
- The reference materials are certified or are the 'best available'.
- The preparation of working standards and reagents is documented.
- Standards, reference materials and reagents are properly labelled and correctly stored.
- New batches of standards and reagents critical to the performance of the method are compared against old batches before use.
- The correct grade of materials is being used in the analyses.
- Where standards, calibrants, or reference materials are certified, copies of the certificate are available for inspection.

6. Quality Control

- There is an appropriate level of quality control for each analysis.
- Where control charts are used, performance has been maintained within acceptable criteria.

(*continued overleaf*)

Table 9.7 (*continued*)

- QC check samples are being tested by the defined procedures, at the required frequency and there is an up-to-date record of the results and actions taken where results have exceeded action limits.
- Results from the random re-analysis of samples show an acceptable measure of agreement with the original analyses.
- Where appropriate, performance in proficiency testing schemes and/or interlaboratory comparisons is satisfactory and has not highlighted any problems or potential problems. Where performance has been unsatisfactory, corrective action has been taken.

7. Sample Management

- There is an effective documented system for receiving samples, identifying samples against requests for analysis, showing progress of analysis, issue of report and fate of sample.
- Samples are properly labelled and stored.

8. Records

- Notebooks/worksheets or other records show the date of analysis, analyst, analyte, sample details, experimental observations, quality control, all rough calculations, any relevant instrument traces and relevant calibration data.
- Notebooks/worksheets are completed in ink and the records are signed or initialled by the analysts.
- Mistakes are crossed out and should not be erased or obliterated. Where a mistake is corrected, the alteration is signed or initialled by the person making the correction.
- The laboratory's procedures for checking data transfers and calculations are being followed.
- Appropriate validation procedures are being followed for electronic records.

9. Test Reports

- The information given in reports is consistent with the requirements of the relevant management standard and reflects any provisions made in the documented method.
- An accreditation body logo is not being used inappropriately.

Table 9.7 *(continued)*

10. Miscellaneous

- There are documented procedures in operation for ascertaining customer satisfaction and handling queries and complaints and system failures.
- The Laboratory Quality Manual is up-to-date and is accessible to all staff.
- There are documented procedures for subcontracting work.
- Vertical audits on random samples (i.e. checks made on a sample, examining all procedures associated with its testing from receipt through to the issue of a report) have not identified any problems.

Appendix

Two-Tailed Critical Values for Student t-Tests

ν	Level of confidence (%)						
	80	90	95	98	99	99.8	99.9
1	3.078	6.314	12.71	31.82	63.66	318.3	636.6
2	1.886	2.920	4.303	6.965	9.925	22.33	31.60
3	1.638	2.353	3.182	4.541	5.841	10.21	12.92
4	1.533	2.132	2.776	3.747	4.604	7.173	8.610
5	1.476	2.015	2.571	3.365	4.032	5.893	6.869
6	1.440	1.943	2.447	3.143	3.707	5.208	5.959
7	1.415	1.895	2.365	2.998	3.499	4.785	5.408
8	1.397	1.860	2.306	2.896	3.355	4.501	5.041
9	1.383	1.833	2.262	2.821	3.250	4.297	4.781
10	1.372	1.812	2.228	2.764	3.169	4.144	4.587
11	1.363	1.796	2.201	2.718	3.106	4.025	4.437
12	1.356	1.782	2.179	2.681	3.055	3.930	4.318
13	1.350	1.771	2.160	2.650	3.012	3.852	4.221
14	1.345	1.761	2.145	2.624	2.977	3.787	4.140
15	1.341	1.753	2.131	2.602	2.947	3.733	4.073
16	1.337	1.746	2.120	2.583	2.921	3.686	4.015
17	1.333	1.740	2.110	2.567	2.898	3.646	3.965
18	1.330	1.734	2.101	2.552	2.878	3.610	3.922

(*continued overleaf*)

Quality Assurance in Analytical Chemistry E. Prichard and V. Barwick

(continued)

	Level of confidence (%)						
ν	80	90	95	98	99	99.8	99.9
19	1.328	1.729	2.093	2.539	2.861	3.579	3.883
20	1.325	1.725	2.086	2.528	2.845	3.552	3.850
21	1.323	1.721	2.080	2.518	2.831	3.527	3.819
22	1.321	1.717	2.074	2.508	2.819	3.505	3.792
23	1.319	1.714	2.069	2.500	2.807	3.485	3.768
24	1.318	1.711	2.064	2.492	2.797	3.467	3.745
25	1.316	1.708	2.060	2.485	2.787	3.450	3.725
26	1.315	1.706	2.056	2.479	2.779	3.435	3.707
27	1.314	1.703	2.052	2.473	2.771	3.421	3.690
28	1.313	1.701	2.048	2.467	2.763	3.408	3.674
29	1.311	1.699	2.045	2.462	2.756	3.396	3.659
30	1.310	1.697	2.042	2.457	2.750	3.385	3.646
40	1.303	1.684	2.021	2.423	2.704	3.307	3.551
50	1.299	1.676	2.009	2.403	2.678	3.261	3.496
60	1.296	1.671	2.000	2.390	2.660	3.232	3.460
70	1.294	1.667	1.994	2.381	2.648	3.211	3.435
80	1.292	1.664	1.990	2.374	2.639	3.195	3.416
90	1.291	1.662	1.987	2.368	2.632	3.183	3.402
100	1.290	1.660	1.984	2.364	2.626	3.174	3.391
∞	1.282	1.645	1.960	2.326	2.576	3.090	3.291

Responses to Self-Assessment Questions

Chapter 3

Response 3.1a

Both (i) and (ii) are false. The errors introduced in sampling cannot be controlled by the use of standards or reference materials.

Response 3.1b

(i) *Representative sample.* The water may arise from ice or snow and the dissolved material may vary. It is the average concentration in the river water which is usually required.

(ii) *Random sample.* For food analysis, there may be a prescribed sampling plan – otherwise use an appropriate random sampling plan where each can has an equal chance of being selected.

(iii) *Selective sample.* In this case, if the contaminant needs to be identified, you do not want it to be diluted and so samples near the point of contamination are selected.

(iv) *Selective sample.* The sacks nearest to the hydrocarbon source would be those most likely to be contaminated and so, in the first instance, these should be examined.

(v) *Representative sample.* If the bags are stored appropriately, there is no reason to believe that the % moisture of different bags would be different. Any bag selected at random would therefore be considered representative of the parent material.

Response 3.2

The first step is to determine the number of bags of peas that need to be selected from the lot at random. Table 3.1 shows that for a lot size of 3000 and inspection Level II the appropriate sample size code letter is 'K'. Table 3.2 shows single sampling plans for normal inspection. This tells you that sample size code letter 'K' equates to a sample size of 125. This is the number of bags of peas that must be selected from the lot at random. The Acceptance Quality Limit (AQL) has been set at 6.5%. Looking down the AQL = 6.5 column and across row K you will see that the acceptance number for this sampling plan is 14. Therefore, as long as the analyst finds no more than 14 bags of peas from the sample of 125 containing in excess of 10 wt% defective peas, the lot is accepted.

Response 3.3

(i) The fish and the brine would normally be analysed separately and not homogenized since the brine is not normally consumed.

(ii) Homogenize, since the fruit and syrup are both eaten.

(iii) Homogenize, since the fruit and natural juice are both consumed.

(iv) Homogenize, since this type of sauce is normally eaten with the fish.

It is important to remember that decisions on the treatment of samples prior to analysis should always be based on sound knowledge of what the results are going to be used for. It is therefore important to establish a good dialogue with the customer prior to carrying out any tests.

Chapter 4

Response 4.1

Your answer will depend to a large extent on a number of assumptions that have to be made regarding the grade of analyst used, level of overheads applied and time taken for each operation. In this question you are given the hourly rate of the analyst but have to estimate the time for each operation. Some suggestions can be found below.

For (a):

(i) Grinding the ore, sieving (to a pre-determined particle size) and removal of test portion for analysis. Dissolution in water and making up to volume. 30 min

(ii) Taking three aliquot portions, addition of KI and titration of the liberated iodine with standard thiosulfate solution. 15 min

(iii) Preparation and standardization of thiosulfate solution. 30 min

(iv) Repetition of the analysis using two other test portions. 30 min

Hence, the total time for the analysis of a single sample would be (i) + (ii) + (iii) + (iv), i.e. 105 min.

If six samples were submitted for analysis, the total time required would be:

$$6 \times [\text{(i)} + \text{(ii)} + \text{(iv)}] + \text{(iii)} = 480\,\text{min or } 80\,\text{min per sample.}$$

If the analyses were to be checked using a reference material or using an alternative technique, similar savings in cost would be obtained on a batch of six samples as opposed to a single sample.

Hence, the total cost for a fully validated analysis could be calculated as follows:

Single sample: 105 min + 105 min (say) for check analysis = 210 min.

Assuming an analyst costs £50 per hour, cost = £175.

Time for six samples would be 480 + 105 = 585 min.

Cost for six samples would therefore be £50 × 585/60 = £487.50 or £81.25 per sample.

For (b):

(i) Grinding the feed, sieving, mixing and removal of a test portion for analysis. 20 min

(ii) Digestion with concentrated acids (intermittent attention), extraction into an organic solvent and determination. 45 min

(iii) Preparation of Cu calibration solutions over a suitable range and extraction into an organic solvent. Preparation of calibration graph. 60 min

(iv) Repetition of analysis using two other portions. Two test portions processed at the same time. 45 min

Hence, the total time required for the analysis of a single sample is (i) + (ii) + (iii) + (iv), i.e. 170 min.

If six samples were submitted, the total time required would be:

$$6 \times \text{(i)} + \text{(iii)} + 3 \times \text{(ii)} + 3 \times \text{(iv)} = 450 \text{ min} \text{ or } 75 \text{ min per sample.}$$

The time for six samples is *not* $6 \times [\text{(i)} + \text{(ii)} + \text{(iv)}]$ since once the grinding is completed the other stages can be carried out on two test portions at the same time.

Cost for one sample = £141.67.

Cost per sample if six analysed = £62.50 (Total cost of £375).

The cost per sample decreases when you have a number of analyses of the same type.

Response 4.2

To calculate the *RSD* value expected for the interlaboratory reproducibility a starting point is to apply the Horwitz function.

The concentration in this case is very low, i.e. $0.562\,\mu\text{g kg}^{-1}$ (below 1.2×10^{-7}) and therefore, according to equation (4.8), the expression to use is:

$$s_R = 0.22c$$

The concentration has to be expressed as a fraction, i.e. $5.62 \times 10^{-10}\,\text{g g}^{-1}$

$$s_R = 0.22 \times 5.62 \times 10^{-10}$$
$$= 1.24 \times 10^{-10}$$

or, in the original units, $0.124\,\mu\text{g kg}^{-1}$.

This is equivalent to a *%CV* of 22%. This is the highest *%CV* that this approach can calculate.

If the original Horwitz function had been used, a larger value would have been obtained. If you check Figure 4.6, that predicts a value for the *%CV* of about 50%. Using the original function, equation (4.4), gives this value.

$$s_R = 0.02 \times c^{0.8495}$$
$$= 0.02 \times (5.62 \times 10^{-10})^{0.8495} = 0.02 \times 1.386 \times 10^{-8}$$
$$= 2.77 \times 10^{-10}\text{g g}^{-1}$$
$$= 0.277\,\mu\text{g kg}^{-1}$$

This is a *%CV* of 49%.

Use of the original equation for low concentrations is now considered to give an underestimate of the laboratory performance [7].

Response 4.3

The results obtained for the analysis given in the question are shown in Table SAQ 4.3. The first step is to calculate the mean value.

Table SAQ 4.3 Results obtained for the analysis of the CRM

Experiment number	Cholesterol (mg $(100\,g)^{-1}$)
1	271.4
2	266.3
3	267.8
4	269.6
5	268.7
6	272.5
7	269.5
8	270.1
9	269.7
10	268.6
11	268.4
mean	269.3

The mean value obtained from the analytical results is 269.3 mg $(100\,g)^{-1}$.

The bias (B) is the difference between the experimental mean value and the value assigned to the CRM.

In this case:

$$B = 269.3 - 274.7 = -5.4\,\text{mg}\ (100\text{g})^{-1}$$

The percentage bias is given by equation (4.10):

$$\%B = \frac{(\overline{x} - x_0)}{x_0} \times 100 = \frac{(269.3 - 274.7)}{274.7} \times 100 = -1.96.$$

The percentage recovery can be calculated by using either equation (4.11) or equation (4.12).

From equation (4.11):

$$\%R = \frac{269.3}{274.7} \times 100 = 98$$

Using equation (4.12):

$$\%R = 100 + \%B = 100 - 1.96 = 98$$

This would not be considered a significant bias especially as the uncertainty in the CRM value is much larger (>3%). To obtain an objective judgement, the uncertainty of both the CRM and of the method should be taken into account and a significance test applied.

Chapter 5

Response 5.1

Of the eight statements, only (b), (c) and (g) are true. The others are false:

(a) Working carefully is likely to minimize mistakes but rarely rules them out altogether – it is always worth doing some checking.

(d) The analyst who has done the work should always be the one to clear up afterwards. Other analysts, apart from not having made the mess in the first place, will not have the same first-hand knowledge of hazards and cleaning requirements.

(e) Samples should normally be allowed to reach room temperature before analysis. Liquid samples, for example, expand as they warm up and so if they are 'sampled cold', an incorrect volume will be taken.

(f) Volumetric glassware should never be dried in an oven. When heated, the glassware expands. When allowed to cool it does not necessarily return to its original volume and thus any volume calibration will become invalid.

(h) In developing a method, the quickest, most reliable way of carrying out the procedures will have been established and documented. It is unlikely that short cuts can be introduced without increasing the uncertainty of the result.

(i) A validated method does not guarantee the correct result – the analyst has to be trained and be able to demonstrate competence to carry out the method.

Response 5.2

(i) The data are plotted in Figure SAQ 5.2 and give a reasonable straight-line plot.

The equation for the line is:

$$y = 2.09x + 5.41$$

The concentrations corresponding to the three responses are, to the nearest whole number:

(a) $5\,\mathrm{mg\ ml^{-1}}$

(b) $15\,\mathrm{mg\ ml^{-1}}$

(c) $24\,\mathrm{mg\ ml^{-1}}$.

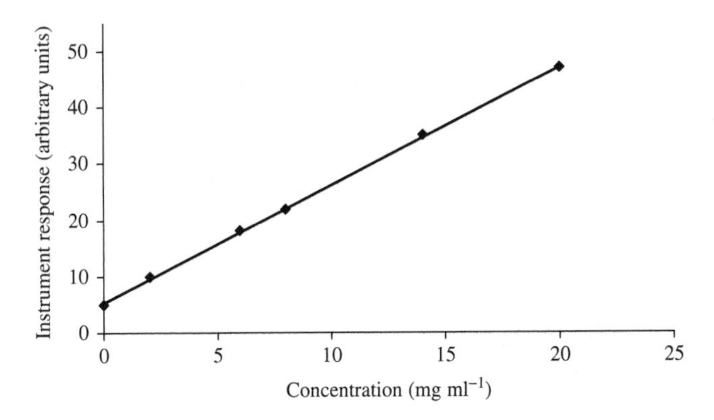

Figure SAQ 5.2 Graph of instrument response against concentration.

Note that in (c) the response corresponds to a point on the graph above the highest concentration of the chemical standard. Although the graph as plotted gives a fairly good straight line you should not assume that this linearity can be extrapolated. Check the linearity by measuring a higher concentration of the chemical standard, say 30 mg ml^{-1}.

(ii) The original concentration of analyte is 0.24 mg ml^{-1}.

Proof:

Y = concentration of the added internal chemical standard = 5 mg ml^{-1}

A = the response of the unknown concentration of the analyte = 5

B = the total response of the unknown concentration of analyte plus added chemical standard = 15

$$C = (\text{Volume}_{\text{Standard}})/(\text{Volume}_{\text{Sample}} + \text{Volume}_{\text{Standard}}) = \frac{1}{(9+1)} = \frac{1}{10}$$

$$D = (\text{Volume}_{\text{Sample}})/(\text{Volume}_{\text{Sample}} + \text{Volume}_{\text{Standard}}) = \frac{9}{(9+1)} = \frac{9}{10}$$

The original concentration of analyte:

$$X = \frac{YAC}{[B - (DA)]} = \frac{5 \times 5 \times 0.1}{[15 - (0.9 \times 5)]} = 0.24 \text{ mg ml}^{-1}$$

The original concentration of analyte in the sample is 0.24 mg ml^{-1}.

Response 5.3

The tabulated data are plotted and the equation of the line obtained. From the slope and the intercept, the initial concentration can be calculated. Alternatively, the graph can be extended until the negative intercept on the x-axis is reached. The graph you obtain should look like the one shown in Figure SAQ 5.3.

The equation of the line is:

$$y = 0.0535x + 2.57$$

The concentration of zinc in the soil sample solution is calculated from the slope and intercept or by projecting the line to the value of x corresponding to $y = 0$. The concentration of zinc in the sample is 48.0 ng ml^{-1}.

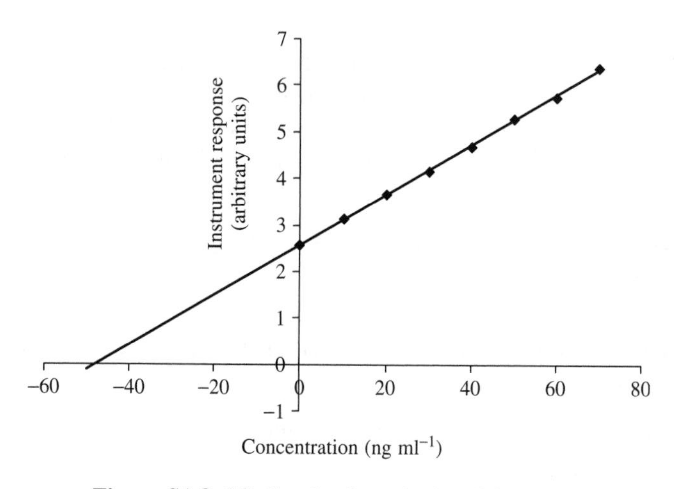

Figure SAQ 5.3 Graph of standard additions data.

Response 5.4

(i) Grade for use to analyse for pesticide residues by gas chromatography.

The main consideration in this case is that the solvent should not interfere with the gas chromatography of the pesticides. Organic impurities will probably be evident as small side peaks on the chromatogram; trace metal contamination is likely to be very low and would not show up in the chromatogram anyway and is probably unimportant in this case. Therefore, the hexane should be as pure as possible; look for a grade with an assay that is of high purity or specifies that it is suitable for your requirements. The 'Distol F' grade n-hexane would probably be appropriate.

(ii) Grade for use as a solvent for oil extraction.

Once the hexane has been used for the oil extraction it will be removed and the oil isolated. Because of the relative boiling points of the hexane and the oil, it will be simple to remove the hexane by distillation. Whatever the grade of hexane used, inorganic impurities are likely to be insignificant and can be discounted. Organic impurities are likely to boil at temperatures close to hexane and can therefore also be removed from the oil by distillation. Thus, the 'Certified AR' grade is suitable.

(iii) Grade for use for HPLC analysis with UV detection.

The main consideration in this case is the transmission property of the hexane, in other words how freely it will let light pass through it. Pure hexane has good transmission in this wavelength range. Impurities in the hexane, in particular, aromatic compounds, will reduce transmission. Choose a grade which has specifically had the interfering impurities removed. Hexane is available in a 'HPLC grade' specifically for UV use. This product may not be suitable for fluorescence detection. The transmission might be adequate but there is no information about its fluorescent properties. You would need to check with the manufacturer for additional information or measure them yourself.

Response 5.5

(i) Label for identifying defective equipment.

The label should be fairly obvious and so will need to be a significant size in relation to the item of equipment. For a small item such as a thermometer, a luggage label would be suitable. For a large instrument, a larger label would be appropriate, perhaps 15×10 cm. In each case, some means of fixing the label in place will be necessary. The label should reference any serial numbers which uniquely identify the instrument or its separate parts. The person under whose authority the label has been issued should be identified by signature or initial, and dated. The message should be brief and clear, e.g. 'OUT OF SERVICE – NOT TO BE USED UNTIL FURTHER NOTICE'.

(ii) Label for identifying a chemical solution of known concentration.

The solution will have been made up in a volumetric flask and ideally should have been transferred to a suitable storage bottle. Storage of these chemical standard solutions in volumetric flasks is a common practice but not one to be encouraged. It is also common practice to write on the storage bottle with a so-called indelible marker. The writing will probably be made unreadable the first time the side of the bottle comes in contact with any liquid; this type of labelling is very vulnerable. Plastic tape labels will not smudge, but spillage of the solution may affect the adhesive so that the tape falls off. Probably the most suitable type of label is either a luggage label firmly attached to the neck of the bottle, perhaps wired on with a 'freezer-tie', or a paper label stuck to the main body of the bottle and protected from spillage by covering with a transparent tape. The label should be written with a non-smudge, water-insoluble ink.

For this particular application the following information should be included on the label:

(a) identity of reagent, and its concentration, details of solvent (with pH, if relevant);

(b) intended use, with restrictions, if relevant;

(c) date prepared, with analyst's initials and date of expiry, if relevant;

(d) hazard warning and special storage needs (e.g. 'keep refrigerated'), if relevant.

(iii) Label for waste solvent drum.

The type of container used for waste solvent will vary according to the type of waste involved, e.g. steel drums for organic solvents and glass bottles for inorganic or corrosive waste. It is important to realize that waste containers are a potential hazard unless care is taken to ensure that particular waste compounds which react vigorously together are not allowed to mix. It may be practical to 'colour-code' containers for different types of waste. However, the information on the label of the container is obviously very important. Since waste is usually put into recycled containers, any existing labels, including hazard warnings, must be removed or obscured, and appropriate new ones substituted. Self-adhesive hazard labels showing internationally recognized hazard symbols are commercially available.

The most practical type of labelling is probably as described in (ii) above. The label should have the following information:

(a) 'WASTE SOLVENT', types of solvent which may be disposed of, types of solvent which may not be disposed of;

(b) date container put into use;

(c) emergency contact;

(d) requirements for washing or neutralization of waste before adding to the container.

Response 5.6

For sample (a), a glass bottle with a screw top is the preferred choice. Assuming it is a clean bottle, the glass will add nothing to the sample which might cause interferences. If organic compounds in the water are likely to be lost to the inside of the bottle by adsorption, a suitable extracting solvent is added to the bottle prior to sampling. The washer to seal the top of the bottle may need to be chosen with care. For some pollutants it may be acceptable to use a polypropylene bottle. For sample (b), probably all of the containers could be suitable. The choice might become more limited if the powder is known to have corrosive properties. The use of bottles might be unwise if the powder is not free flowing. For sample (c), the only really unsuitable container is the polyethylene bag. The bottles and the can are probably all suitable, although the can is probably preferable if the sample is to be stored for any length of time, particularly if the dyestuff is liable to fade with prolonged exposure to light. If glass has to be used, then a brown glass bottle would reduce the problem with light exposure.

Chapter 6

Response 6.1

Mean

$$\overline{x} = \frac{\sum_{i=1}^{n} x_i}{n} = 269.3\,\text{mg}\ (100\,\text{g})^{-1}$$

Sample standard deviation

$$s = \sqrt{\frac{\sum_{i=1}^{n}(x_i - \overline{x})^2}{n-1}} = 1.69\,\text{mg}\ (100\,\text{g})^{-1}$$

Relative standard deviation

$$RSD = \frac{s}{\overline{x}} = 0.0063$$

Degrees of freedom

$$\nu = n - 1 = 10$$

95% confidence interval for the mean

$$\overline{x} \pm t_{(\nu,\alpha)} \times \frac{s}{\sqrt{n}}$$

$$\pm\ 2.228 \times \frac{1.69}{\sqrt{11}} = 1.14\,\text{mg}\ (100\,\text{g})^{-1}$$

The Student t-value for 10 degrees of freedom at the 95% confidence level is 2.228.

Response 6.2

(a) The Shewhart chart, with the appropriate limits (see (d)) is shown in Figure SAQ 6.2(a).

(b) The moving average chart ($n = 5$), with the appropriate limits (see (e)) is shown in Figure SAQ 6.2(b), while the data required to construct the chart are shown in Table SAQ 6.2(a).

(c) The CUSUM chart is shown in Figure SAQ 6.2(c), while the data required to construct the chart are shown in Table SAQ 6.2(b).

Table SAQ 6.2(a) Data for moving average chart

Data	16	16	18	14	16	15	18	17	18
Moving average ($n = 5$)	—	—	—	—	16.0	15.8	16.2	16.0	16.8
Data	18	16	18	15	16	17	21	17	21
Moving average ($n = 5$)	17.2	17.4	17.4	17.0	16.6	16.4	17.4	17.2	18.4
Data	20	22	19	19	21	22	20	21	20
Moving average ($n = 5$)	19.2	20.2	19.8	20.2	20.2	20.6	20.2	20.6	20.8
Data	19	22	21	21	21	22	21	21	—
Moving average ($n = 5$)	20.4	20.4	20.6	20.6	20.8	21.4	21.2	21.2	—

Table SAQ 6.2(b) Data for CUSUM chart

Data	16	16	18	14	16	15	18	17	18
Data − target	−1	−1	1	−3	−1	−2	1	0	1
CUSUM	−1	−2	−1	−4	−5	−7	−6	−6	−5
Data	18	16	18	15	16	17	21	17	21
Data − target	1	−1	1	−2	−1	0	4	0	4
CUSUM	−4	−5	−4	−6	−7	−7	−3	−3	1
Data	20	22	19	19	21	22	20	21	20
Data − target	3	5	2	2	4	5	3	4	3
CUSUM	4	9	11	13	17	22	25	29	32
Data	19	22	21	21	21	22	21	21	—
Data − target	2	5	4	4	4	5	4	4	—
CUSUM	34	39	43	47	51	56	60	64	—

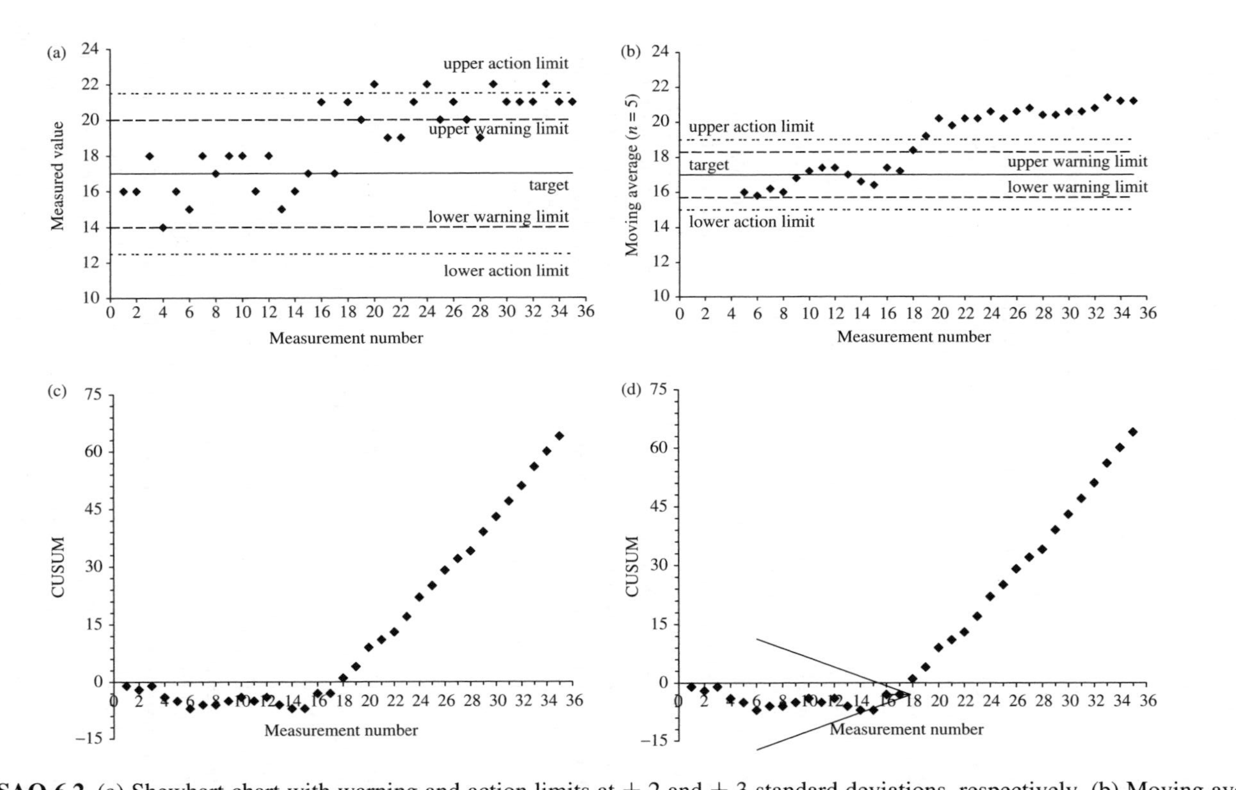

Figure SAQ 6.2 (a) Shewhart chart with warning and action limits at $\pm$ 2 and $\pm$ 3 standard deviations, respectively. (b) Moving average chart ($n = 5$) with warning and action limits at $\pm\ 2\sqrt{n}$ and $\pm\ 3\sqrt{n}$ standard deviations, respectively. (c) CUSUM chart. (d) CUSUM chart with V-mask.

(d) Warning limits should be plotted at 14 and 20 and action limits should be plotted at 12.5 and 21.5, as shown in Figure SAQ 6.2(a). The system appears to go 'out of control' at measurement numbers 18 to 19. Measurement 18 is between the warning and action limits, measurement 19 is right on the warning limit and measurement 20 is outside the action limit. All three points are on the same side of the mean.

(e) Warning limits should be plotted at 15.7 and 18.3 and action limits should be plotted at 15 and 19, as shown in Figure SAQ 6.2(b). Again, the system appears to go 'out of control' around measurement numbers 18 to 19. Measurement 18 is between the warning and action limits while measurement 19 is outside the action limit. Note the smoothing action of the averaging. The step-change in the data is visually more obvious compared to the Shewhart chart.

(f) On the CUSUM chart, the data change direction very obviously at measurement 16. The change is already well-established by measurements 18 to 19. When setting up the CUSUM chart, the divisions on both the y- and x-axes should be the same length. A division on the x-axis should represent a single unit, while a division on the y-axis should be equivalent to 2σ. In this case, the divisions on the y-axis should represent 3 units of measurement (see Figure SAQ 6.2(c). A 'V-mask' should be constructed so that the length of d is equal to two divisions of the x-axis. The angle θ should be approximately 22° and the total length of the mask in the horizontal direction should be equal to 12 divisions of the x-axis. Using a mask with these dimensions will indicate that the system goes 'out of control' at measurement 16, as shown in Figure SAQ 6.2(d).

Response 6.3

Uncertainty is a parameter which characterizes the range of values within which the value of the quantity being measured is expected to lie.

Therefore, the correct answer is (d).

Answer (a) is wrong because error is the difference between an individual result and the true value of the quantity being measured. It is expressed as a single value, whereas uncertainty characterizes a range of values.

If you answered (b), perhaps you were thinking of the spread of values obtained from replicate measurements. While these do indeed form a range, one such range will relate to only one source of uncertainty and there may be several sources of uncertainty affecting a particular measurement. The precision of a measurement is an indication of the random error associated with it. This takes no account of any systematic errors that may be connected with the measurement. It is important to realize that uncertainty covers the effects of both random error and systematic error and, moreover, takes into account multiple sources of these effects where they are known to exist and are considered significant.

Answer (c) is incorrect as accuracy and uncertainty are not related in the way suggested.

If you thought that (e) was correct, look again at the definition of uncertainty. We would expect the true value to lie within the range of values defined by the uncertainty and not to form one of the boundaries of that range.

Answer (f) is incorrect as uncertainty is not a probability but a range of values. You should note, however, that a probability, in the form of a confidence level, can, and should, be assigned to an uncertainty.

Response 6.4

(a) This is a stated tolerance for volumetric glassware and so a rectangular or a triangular distribution is assumed. The standard uncertainty is 0.046 ml ($0.08/\sqrt{3}$) if a rectangular distribution is assumed, or 0.033 ml ($0.08/\sqrt{6}$) if a triangular distribution is assumed.

(b) The value given is an expanded uncertainty. The latter is converted to a standard uncertainty by dividing by the stated coverage factor, in this case, $k = 2$. The standard uncertainty is therefore 0.0002 g.

(c) The uncertainty quoted with the purity value is a stated range. A rectangular distribution is generally assumed in such cases, as we have no reason to believe that the actual purity will be closer to 99.9% than 99.8% or 100%. The standard uncertainty is therefore $0.1/\sqrt{3} = 0.058\%$.

(d) In this case, the data are expressed as a standard deviation. A standard uncertainty is an uncertainty estimate expressed as a standard deviation and so no conversion is necessary.

Response 6.5

Remember that the mathematical model we are dealing with is of the form:

$$y = abc \qquad \text{or} \qquad y = \frac{a}{b}$$

This means that what we need to combine are ***relative*** standard uncertainties and not just plain standard uncertainties. The result of combining relative standard uncertainties will itself be a relative standard uncertainty.

The correct equation is (d). At first glance it may not appear that (d) produces a relative standard uncertainty, but look again. We can rewrite (d) as follows:

$$\frac{u(y)}{y} = \sqrt{\left(\frac{u(a)}{a}\right)^2 + \left(\frac{u(b)}{b}\right)^2 + \left(\frac{u(c)}{c}\right)^2}$$

The left-hand side of this equation now clearly represents a relative standard uncertainty. If we multiply both sides of this equation by y it is transformed into (d).

Equation (a) is incorrect because y does not appear on one side of the equation.

Equation (b) is wrong because it actually represents the inverse of the combined relative standard uncertainty.

Equation (c) is not correct because the squares of the component relative standard uncertainties have been multiplied together instead of being added.

Equation (e) is incorrect because the reciprocals of the component relative standard uncertainties have been combined.

Well done if you got this one right. If you picked the wrong answer, do not be too disappointed because, as you can see, it is very easy to get equations 'mixed-up'. Go over the section again and try making up some of your own exercises to test the rules of uncertainty combination. You are bound to improve with practice.

Response 6.6

This example addresses the quantification and combination stages of uncertainty calculations. The first stage is to convert any data that are not already given as a standard uncertainty (i.e. prefixed by u). In this example, the data associated with the purity ($\pm$ 0.1%) and the information from the supplier's catalogue on the volume of the solution ($\pm$ 0.4 ml) fall into this category.

In the case of the purity, we have no reason to suppose that any value between the quoted extremes of $-$ 0.1% and $+$ 0.1% is more likely than any other. This being the case, we should assume a rectangular distribution and our standard uncertainty for purity is therefore:

$$u(P) = 0.1/\sqrt{3} = 0.057\ 735$$

The value quoted in the supplier's catalogue for the 1000 ml volumetric flask is a manufacturing tolerance. In this case, you could assume a rectangular or a triangular distribution. For a more conservative, initial estimate of the uncertainty you can assume a rectangular distribution (if you chose a triangular distribution, this is not incorrect, but it will give you a slightly smaller combined uncertainty):

$$u_2(V) = 0.4/\sqrt{3} = 0.230\ 94\ \text{ml}$$

We have a second standard uncertainty for the volume of the volumetric flask. This estimate was obtained by making replicate measurements of the volume. The two standard uncertainties relating to the volume must be combined to produce a single value. This is achieved by a straightforward application of equation (6.12):

$$u(V) = \sqrt{0.1^2 + 0.230\ 94^2} = 0.251\ 66\ \text{ml}$$

There are two standard uncertainties associated with the mass of the KHP used and these can be combined in a similar fashion:

$$u(M) = \sqrt{0.000\ 07^2 + 0.000\ 05^2} = 0.000\ 086\ 023\ \text{g}$$

We have now reached the stage where we have a single value for the uncertainty connected with each parameter in the model. It is worth collecting these values together. Note that we have retained a large number of significant figures at this stage – the result will be rounded once the expanded uncertainty has been calculated.

$$u(M) = 0.000\ 086\ 023 \text{ g}$$

$$u(P) = 0.057\ 735$$

$$u(V) = 0.251\ 66 \text{ ml}$$

$$u(F) = 0.0017 \text{ g mol}^{-1}$$

The next step is to calculate the concentration of the KHP solution. From the model, this is given by:

$$C = \frac{1000 \times 20.4220 \times 99.9}{1000 \times 204.2236 \times 100} = 0.099\ 898 \text{ mol l}^{-1}$$

The model for C consists of a mixture of multiplication and division and so the standard uncertainty of C is obtained by an application of equation (6.13):

$$u(C) = 0.099\ 898 \times \sqrt{\left(\frac{0.000\ 086\ 023}{20.4220}\right)^2 + \left(\frac{0.057\ 735}{99.9}\right)^2 + \left(\frac{0.251\ 66}{1000}\right)^2 + \left(\frac{0.0017}{204.2236}\right)^2}$$

$$= 0.000\ 062\ 977 \text{ mol l}^{-1}$$

As we have been asked to report the figure we have found with a 95% confidence level, we must now multiply the standard uncertainty calculated above by a coverage factor of 2. This produces:

$$0.000\ 062\ 977 \times 2 = 0.000\ 125\ 95 \text{ mol l}^{-1}$$

Having retained as many figures as possible in our calculations, the final step is to 'round' the values to a level commensurate with the input data. Thus, what we actually report is:

$$\text{Concentration of KHP} = 0.0999 \pm 0.0001 \text{ mol l}^{-1}$$

where the reported uncertainty is an expanded uncertainty calculated using a coverage factor of 2, which gives a level of confidence of approximately 95%.

Did you remember to multiply the standard uncertainty for the concentration by the coverage factor? Remember that you only apply the coverage factor at the very end of your calculations when you wish to associate a confidence level with the uncertainty value you are reporting. A standard uncertainty on its own has a confidence level of approximately 68% but this would be too low a level for practical use. If you wanted to be even more confident of the accuracy of your measurement, you could use a coverage factor of 3 – this would give you a confidence level of 99.7%. You were not asked to do this here. If you had been, the expanded uncertainty would have been 0.000 19. So, the more confident you want to be in your expression of a measurement value, the wider must be the band of uncertainty surrounding it. Although this may sound like a contradiction at first, if you think about it you will see that it makes sense.

Chapter 7

Response 7.1

(a) The mean, standard deviation, robust average (median) and robust standard deviation (MAD_E) are shown in Table SAQ 7.1.

Table SAQ 7.1 Data from a proficiency testing round for the determination of moisture in barley

Laboratory identity number	Result (wt%)	$\lvert x_i - \text{median}\rvert$	$z = \dfrac{x_i - X}{\hat{\sigma}}$
1	13.4	0.0	0.0
2	13.5	0.1	0.5
3	13.4	0.0	0.0
4	13.2	0.2	−1.0
5	13.6	0.2	1.0
6	12.7	0.7	−3.5
7	13.3	0.1	−0.5
8	13.6	0.2	1.0
9	13.6	0.2	1.0
10	13.4	0.0	0.0
11	13.2	0.2	−1.0
12	13.7	0.3	1.5
13	13.4	0.0	0.0
14	13.3	0.1	−0.5
15	13.7	0.3	1.5
16	13.2	0.2	−1.0
17	13.3	0.1	−0.5
mean	13.4		
standard deviation	0.24		
median	13.4		
	MAD	0.2	
	MAD_E	0.3	

To calculate the robust standard deviation for this data set, you first have to calculate the absolute difference between each result and the median, and then find the median of these values. The median absolute deviation (MAD) is 0.2 wt%. This is converted to a standard deviation equivalent (MAD_E) by multiplying by 1.483:
Robust standard deviation, $MAD_E = 1.483 \times 0.2 = 0.3$ wt%.

(b) In general, the robust average (median) should be used as the assigned value. This is a more reliable estimate as the influence of any extreme values in the data set is reduced. In this example, the robust estimates of the mean and standard deviation are not very different from the normal statistical estimates as there are no really extreme values in the data set.
The calculated z-scores for each laboratory are shown in Table SAQ 7.1. These were calculated by using an assigned value of 13.4 wt% and a target range of 0.2 wt%.

(c) Laboratory 6 would consider its performance unsatisfactory as it has obtained a z-score $>|3|$. All of the other participants have obtained z-scores of $<|2|$ and so their performance can be considered as being satisfactory.

Chapter 9

Response 9.1

(i) The ISO 9001:2000 Standard applies to the overall operation of an organization, which can include analytical testing as an aspect of production or the provision of a service. However, in the scenario described in the question, what is being looked for is a Standard to form a basis for the management of quality in the analytical laboratory. It would therefore be more appropriate to choose one of the Standards directly concerned with the scientific aspects of the work, that is, either GLP or ISO/IEC 17025.

(ii) The ISO/IEC Standard deals directly with the laboratory's ability to conduct a certain type of analysis/test – it is checking the competence of a laboratory to carry out particular analyses/tests. It would therefore be a very suitable standard to use as a basis for your laboratory's management system in the case described. However, in certain circumstances, there might be other factors which would lead you to choose GLP instead (see (iii) below).

(iii) The GLP principles are intended for use when the results of analytical tests are to be used as part of a regulatory study. You therefore need to know what is going to happen to the data that your laboratory is generating. If, for example, the work formed part of a study to be used to support a request for approval to use a new pesticide on food products, compliance with the GLP principles would be the preferable option to take. In the question, the safety of the pesticide has already been established and its approval for use granted, and your experimental results are to be used for checking that residues are below an agreed level and so accreditation to ISO/IEC 17025 is the more appropriate approach.

Response 9.2

The characteristics to look for in an auditor are difficult to list. It is often easier to identify individuals who would be unsuitable than to be sure a person will be a good auditor. Auditors should be persons of the highest integrity. They require sufficient technical knowledge to be able to assess the scientific aspects of the work being audited. In addition, a thorough knowledge of the Standard against which the tasks are being assessed is required. The person has to have a friendly, approachable manner and is able to focus on the task in hand and not be distracted. Another useful characteristic is the ability to make an objective judgement about the work and to discuss the matter with the auditee without becoming aggressive. Someone who is constantly trying to 'score points' off others will not make a good auditor.

Bibliography[†]

Crosby, N. and Patel, I., *General Principles of Good Sampling Practice*, ISBN 0-85404-412-4, The Royal Society of Chemistry, Cambridge, UK, 1995.

Lawn, R., Thompson, M. and Walker, R. F., *Proficiency Testing in Analytical Chemistry*, ISBN 0-85404-432-9, The Royal Society of Chemistry, Cambridge, UK, 1997.

Huber, L., *Validation and Qualification in Analytical Laboratories*, ISBN 1-57491-080-9, Interpharm Press, Buffalo Grove, IL, USA, 1998.

Baiulescu, G., Stefan, R.-I. and Aboul-Enein, H. Y., *Quality and Reliability in Analytical Chemistry* (Analytical Chemistry Series), ISBN 0-84932-3762, Interpharm Press/CRC Press, Boca Raton, FL, USA, 2000.

Burgess, C., *Valid Analytical Methods and Procedures*, ISBN 0-85404-482-5, The Royal Society of Chemistry, Cambridge, UK, 2000.

Garfield, F., Klesta, M. E. and Hirsch, J., *Quality Assurance Principles for Analytical Laboratories*, 3rd Edition, ISBN 0-935584-70-6, Association of Official Analytical Chemists (AOAC) International, Gaithersburg, MD, USA, 2000.

Eurachem/CITAC, *Quantifying Uncertainty in Analytical Measurement*, Eurachem/CITAC Guide, 2nd Edition, ISBN 0-948926-15-5, (Eurachem), Co-operation on International Traceability in Analytical Chemistry (CITAC), 2000. [http://www.eurachem.org] (accessed 11 December, 2006).

Kenkel, J., *A Primer on Quality in the Analytical Laboratory*, ISBN 1-566-70516-9, Lewis Publishers, Boca Raton, FL, USA, 2000.

Prichard, E., *Analytical Measurement Terminology*, ISBN 0-85404-443-4, The Royal Society of Chemistry, Cambridge, UK, 2001.

LGC, *In-House Method Validation. A Guide for Chemical Laboratories*, ISBN 0-94892-618-X, LGC, Teddington, UK, 2003.

Mullins, E., *Statistics for the Quality Control Chemistry Laboratory*, ISBN 0-85404-671-2, The Royal Society of Chemistry, Cambridge, UK, 2003.

Chan, C. C., Lee, Y. C., Lam, H. and Zhang, X.-M. (Eds), *Analytical Method Validation and Instrument Performance Verification*, ISBN 0-471-25953-5, John Wiley & Sons, Inc., Hoboken, NJ, USA, 2004.

Taylor, J. K. and Cihon, C., *Statistical Techniques for Data Analysis*, 2nd Edition, ISBN 1-58488-385-5, Chapman & Hall/CRC Press, Boca Raton, FL, USA, 2004.

[†]Arranged in chronological order.

Wenclawiak, B., Koch, M. and Hadjicostas, E., *Quality Assurance in Analytical Chemistry*, ISBN 3-54040-57-8, Springer-Verlag, Berlin, Germany, 2004.
Funk, W., Dammann, V. and Donnevert, G., *Quality Assurance in the Analytical Laboratory*, ISBN 3-527-31114-9, Wiley-VCH, Weinheim, Germany, 2006.
Hibbert, D. B., *Quality Assurance in the Analytical Chemistry Laboratory*, ISBN 978-0-19-516212-7 (Hardback) ISBN 978-0-19-516213-4 (Paperback), Oxford University Press Inc., 198 Madison Avenue, NY, USA, 2007.

Glossary of Terms

This section contains a glossary of terms, all of which are used in the text. It is not intended to be exhaustive, but to explain briefly those terms which often cause difficulties or may be confusing to the inexperienced reader.

Accreditation Third-party statement based on a decision following review that competence to carry out a task has been demonstrated.

Accuracy The difference between a single test result and the accepted reference value.

Analyte The substance subject to analysis.

ANOVA (Analysis of Variance) A statistical procedure that is used (a) to test for significant differences among means of several sets of results or (b) to estimate variances of several influences operating independently.

Audit A systematic, independent and documented process for obtaining evidence and evaluating it objectively to determine the extent to which set procedures or requirements have been fulfilled.

Bias The difference between the expectation of the test results and an accepted reference value.

Calibration Operation that establishes the relationship, obtained by reference to one or more measurement standards, between the response of an instrument and the values of the standards.

Certification Third-party statement based on a decision following review that fulfilment of specified requirements have been demonstrated. Usually related to products, processes, systems or persons.

Quality Assurance in Analytical Chemistry E. Prichard and V. Barwick

Certified Reference Material A reference material characterized by a metrologically valid procedure for one or more specified properties, accompanied by a certificate that states the value of the specified property, its associated uncertainty and a statement of metrological traceability.

Coefficient of variation The ratio of the standard deviation to the mean often expressed as a percentage (*see also* Relative standard deviation).

Confidence interval The range about the mean within which a stated percentage of values would be expected to lie. For example, for a normal distribution, approximately 95% of values lie between $\pm\ 2s$.

Control charts Routine charting of data obtained from the analysis of quality control materials to check that the results lie within predetermined limits.

Error The difference between a result of a measurement and the true value of the measurand.

Expanded uncertainty A quantity defining an interval about the result of a measurement that may be expected to encompass a large fraction of the distribution of values that could reasonably be attributed to the measurand.

External audit Audits conducted by external independent organizations (third-party audits) or by persons having an interest in the organization, e.g. customers (second-party audits).

Fitness for purpose A formal process of assessing that a method is suitable for a given application.

Internal audit Audits conducted by the organization itself for management review purposes – also called first-party audits.

Limit of detection The lowest amount of an analyte that can be measured with reasonable statistical certainty.

Limit of quantitation The lowest concentration of an analyte that can be determined with an acceptable level of uncertainty under the stated conditions of the test.

Linearity The ability of the method to obtain test results proportional to the concentration of the analyte.

Matrix All components of the test sample, excluding the analyte.

Mean The (arithmetic) mean of a set of values.

Measurand A particular quantity subject to measurement.

Measurement uncertainty A parameter, associated with the result of a measurement that characterizes the dispersion of the values that could reasonably be attributed to the measurand. (In the broadest sense, uncertainty of measurement means the doubt about the validity of the result of a measurement.)

Precision The closeness of agreement between independent test results obtained under stipulated conditions. NOTES: (i) Precision depends only on the distribution of random errors and does not relate to the true value or the specified value. (ii) The measure of precision is usually expressed in terms of imprecision and computed as a standard deviation of the test results. Less precision is reflected by a larger standard deviation. (iii) 'Independent test results' means results obtained in a manner not influenced by any previous results on the same or similar test object. (iv) Quantitative measures of precision depend critically on the stipulated conditions. Repeatability and reproducibility conditions are particular sets of extreme stipulated conditions.

Predicate rules The underlying requirements set in the Federal Food, Drug and Cosmetic Act and the Public Health Service Act and FDA regulations (e.g. GLP), other than Part 11.

Primary method A method having the highest metrological qualities, whose operations can be completely described and understood, and for which a complete uncertainty statement can be written down in terms of SI units.

Proficiency Testing Determination of laboratory testing performance by means of interlaboratory comparisons.

Quality Assurance Part of quality management focused on providing confidence that quality requirements will be fulfilled.

Quality Control Part of quality management focused on fulfilling quality requirements.

Quality control material A material that is fully characterized in-house or by a third-party, similar in composition to the types of samples normally examined, stable, homogeneous and available in large quantities so that it can be used over a long period of time for monitoring method performance.

Random error Errors that cause results to differ in an unpredictable way.

Recovery If a known amount of analyte is added to a test sample and the test sample is then analysed for that analyte by a particular method, the recovery is that fraction of the amount of analyte added which is found by the method.

Reference Material A material, sufficiently homogeneous and stable with respect to one or more properties, which has been established to be fit for its intended purpose.

Relative standard deviation The ratio of the standard deviation to the mean value (*see also* Coefficient of variation).

Repeatability Precision under repeatability conditions, i.e. conditions where independent test results are obtained with the same method on identical test items in the same laboratory by the same operator using the same equipment within short intervals of time.

Repeatability limit The value less than or equal to which the absolute difference between two test results obtained under repeatability conditions may be expected to be with a probability of 95%.

Reproducibility Precision under reproducibility conditions, i.e. conditions where independent test results are obtained with the same method on identical test items in different laboratories by different operators using different equipment.

Reproducibility limit The value less than or equal to which the absolute difference between two test results obtained under reproducibility conditions may be expected to be with a probability of 95%.

Residual Difference between the observed response and that predicted by a calibration function.

Ruggedness Test of the extent to which the results of an analytical procedure are affected by slight changes in the procedure.

Selectivity The extent to which a method can be used to determine particular analytes in mixtures or matrices without interference from other components of similar behaviour.

Sensitivity The change in the response of a measuring instrument divided by the corresponding change in the stimulus.

Spiked sample A sample prepared by adding a known quantity of analyte to a matrix which is as close to or identical to that of the sample of interest.

Standard deviation A measure of the dispersion of a set of values.

Standard deviation of the mean A measure of the dispersion of a set of mean values.

Standard error of prediction Standard deviation of the predicted value obtained from linear regression.

Standard Operating Procedure A detailed written instruction to achieve uniformity in the performance of a specific function.

Standard uncertainty Uncertainty of a measurement expressed as a standard deviation.

Student *t*-test A statistical test to establish if there is a significant difference between two mean values, taking account of the uncertainties associated with both values.

Systematic error Errors that cause results to differ from the expected result in a predictable way, either always higher or always lower.

Traceability Property of a measurement result whereby the result can be related to a reference through a documented unbroken chain of calibrations, each contributing to the measurement uncertainty.

Trueness Closeness of agreement between the average value obtained from a large series of test results and an accepted reference value. Trueness is normally expressed in terms of bias.

Validation Confirmation by examination and provision of objective evidence that the particular requirements for a specified intended use are fulfilled.

Variance The square of the standard deviation.

Verification Confirmation that the method performance parameters established during method validation can be met.

Within-laboratory reproducibility/intermediate precision Precision under conditions where independent test results are obtained with the same method on identical test items in the same laboratory by different operators using different equipment on different days.

Working range The interval between the upper and lower concentration (amounts) of analyte in the sample (including these concentrations) for which it has been demonstrated that the analytical procedure has a suitable level of uncertainty.

SI Units and Physical Constants

SI Units

The SI system of units is generally used throughout this book. It should be noted, however, that according to present practice, there are some exceptions to this, for example, wavenumber (cm^{-1}) and ionization energy (eV).

Base SI units and physical quantities

Quantity	Symbol	SI Unit	Symbol
length	l	metre	m
mass	m	kilogram	kg
time	t	second	s
electric current	I	ampere	A
thermodynamic temperature	T	kelvin	K
amount of substance	n	mole	mol
luminous intensity	I_v	candela	cd

Prefixes used for SI units

Factor	Prefix	Symbol
10^{21}	zetta	Z
10^{18}	exa	E
10^{15}	peta	P
10^{12}	tera	T

(*continued overleaf*)

Prefixes used for SI units (*continued*)

Factor	Prefix	Symbol
10^9	giga	G
10^6	mega	M
10^3	kilo	k
10^2	hecto	h
10	deca	da
10^{-1}	deci	d
10^{-2}	centi	c
10^{-3}	milli	m
10^{-6}	micro	μ
10^{-9}	nano	n
10^{-12}	pico	p
10^{-15}	femto	f
10^{-18}	atto	a
10^{-21}	zepto	z

Derived SI units with special names and symbols

Physical quantity	SI unit		Expression in terms of base or derived SI units
	Name	Symbol	
frequency	hertz	Hz	$1\ Hz = 1\ s^{-1}$
force	newton	N	$1\ N = 1\ kg\ m\ s^{-2}$
pressure; stress	pascal	Pa	$1\ Pa = 1\ Nm^{-2}$
energy; work; quantity of heat	joule	J	$1\ J = 1\ Nm$
power	watt	W	$1\ W = 1\ J\ s^{-1}$
electric charge; quantity of electricity	coulomb	C	$1\ C = 1\ A\ s$
electric potential; potential difference; electromotive force; tension	volt	V	$1\ V = 1\ J\ C^{-1}$
electric capacitance	farad	F	$1\ F = 1\ C\ V^{-1}$
electric resistance	ohm	Ω	$1\ \Omega = 1\ V\ A^{-1}$
electric conductance	siemens	S	$1\ S = 1\ \Omega^{-1}$
magnetic flux; flux of magnetic induction	weber	Wb	$1\ Wb = 1\ V\ s$
magnetic flux density;	tesla	T	$1\ T = 1\ Wb\ m^{-2}$
magnetic induction inductance	henry	H	$1\ H = 1\ Wb\ A^{-1}$
Celsius temperature	degree Celsius	°C	$1°C = 1\ K$
luminous flux	lumen	lm	$1\ lm = 1\ cd\ sr$

Derived SI units with special names and symbols (*continued*)

Physical quantity	SI unit		Expression in terms of base or derived SI units
	Name	Symbol	
illuminance	lux	lx	$1\ \text{lx} = 1\ \text{lm m}^{-2}$
activity (of a radionuclide)	becquerel	Bq	$1\ \text{Bq} = 1\ \text{s}^{-1}$
absorbed dose; specific energy	gray	Gy	$1\ \text{Gy} = 1\ \text{J kg}^{-1}$
dose equivalent	sievert	Sv	$1\ \text{Sv} = 1\ \text{J kg}^{-1}$
plane angle	radian	rad	1[a]
solid angle	steradian	sr	1[a]

[a] rad and sr may be included or omitted in expressions for the derived units.

Physical Constants

Recommended value of selected constants[a]

Constant	Symbol	Value
acceleration of free fall (acceleration due to gravity)	g_n	$9.806\,65\ \text{m s}^{-2}$ [b]
atomic mass constant (unified atomic mass unit)	m_u	$1.660\,538\,782(83) \times 10^{-27}$ kg
Avogadro constant	L, N_A	$6.022\,141\,79(30) \times 10^{23}\ \text{mol}^{-1}$
Boltzmann constant	k	$1.380\,650\,4(24) \times 10^{-23}\ \text{J K}^{-1}$
electronic charge to mass quotient	$-e/m_e$	$-1.758\,820\,150(44) \times 10^{11}\ \text{C kg}^{-1}$
elementary charge (electron charge)	e	$1.602\,176\,487(40) \times 10^{-19}$ C
Faraday constant	F	$96\,485.3399(24)\ \text{C mol}^{-1}$
ice-point temperature	T_{ice}	273.15 K[b]
molar gas constant	R	$8.314\,472(15)\ \text{J mol}^{-1}\ \text{K}^{-1}$
molar volume of ideal gas (at $T = 273.15$ K and $p = 101.325$ kPa)	V_m	$22.413\,996(39) \times 10^{-3}\ \text{m}^3\ \text{mol}^{-1}$
Planck constant	h	$6.626\,068\,96(33) \times 10^{-34}$ J s
speed of light in vacuum	c	$299\,792\,458\ \text{m s}^{-1}$ [b]
standard atmosphere	atm	101 325 Pa[b]

[a] A concise form of the values is shown, where it is understood that the number in parentheses is the numerical value of the standard uncertainty referred to the corresponding last digits of the quoted result. (For example, $F = 96\,485.3399(24)\ \text{C mol}^{-1}$ could also be written as $F = (96\,485.3399 \pm 0.0024)\ \text{C mol}^{-1}$.)
[b] Exactly defined values.

The Periodic Table

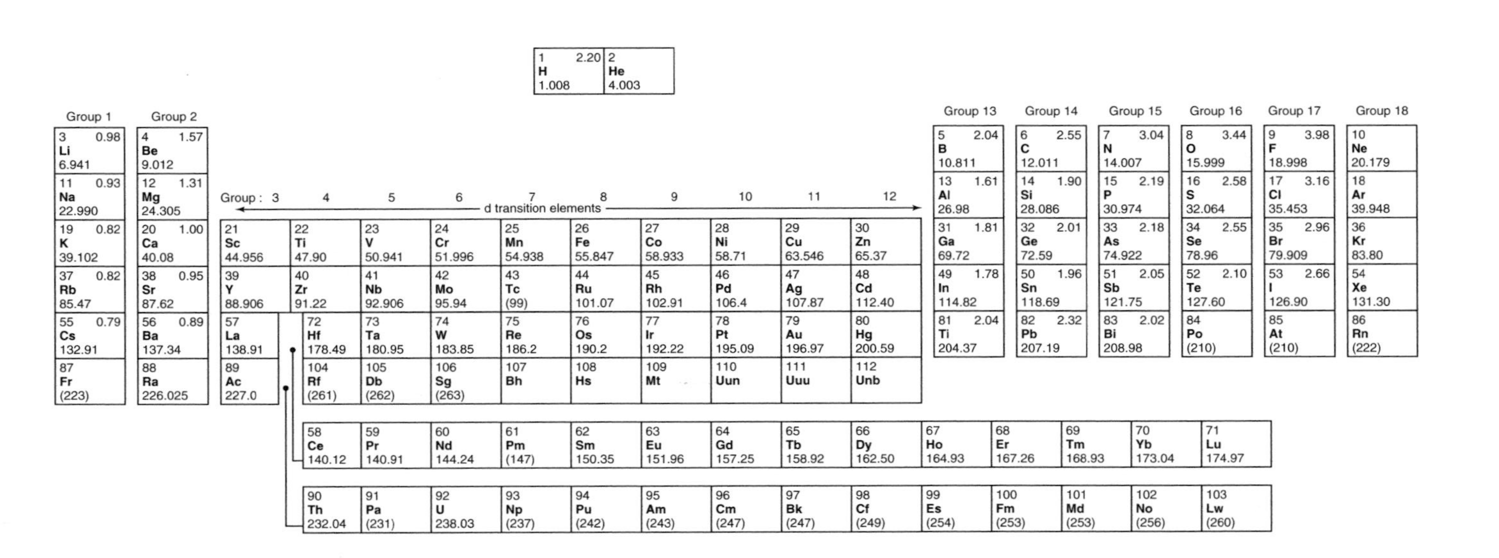

Key	
0.98	Pauling electronegativity
3	Atomic number
Li	Element
6.941	Atomic mass (^{12}C)

1 2.20 **H** 1.008	2 **He** 4.003

Group 1	Group 2	3	4	5	6	7	8	9	10	11	12	Group 13	Group 14	Group 15	Group 16	Group 17	Group 18
3 0.98 **Li** 6.941	4 1.57 **Be** 9.012											5 2.04 **B** 10.811	6 2.55 **C** 12.011	7 3.04 **N** 14.007	8 3.44 **O** 15.999	9 3.98 **F** 18.998	10 **Ne** 20.179
11 0.93 **Na** 22.990	12 1.31 **Mg** 24.305											13 1.61 **Al** 26.98	14 1.90 **Si** 28.086	15 2.19 **P** 30.974	16 2.58 **S** 32.064	17 3.16 **Cl** 35.453	18 **Ar** 39.948
19 0.82 **K** 39.102	20 1.00 **Ca** 40.08	21 **Sc** 44.956	22 **Ti** 47.90	23 **V** 50.941	24 **Cr** 51.996	25 **Mn** 54.938	26 **Fe** 55.847	27 **Co** 58.933	28 **Ni** 58.71	29 **Cu** 63.546	30 **Zn** 65.37	31 1.81 **Ga** 69.72	32 2.01 **Ge** 72.59	33 2.18 **As** 74.922	34 2.55 **Se** 78.96	35 2.96 **Br** 79.909	36 **Kr** 83.80
37 0.82 **Rb** 85.47	38 0.95 **Sr** 87.62	39 **Y** 88.906	40 **Zr** 91.22	41 **Nb** 92.906	42 **Mo** 95.94	43 **Tc** (99)	44 **Ru** 101.07	45 **Rh** 102.91	46 **Pd** 106.4	47 **Ag** 107.87	48 **Cd** 112.40	49 1.78 **In** 114.82	50 1.96 **Sn** 118.69	51 2.05 **Sb** 121.75	52 2.10 **Te** 127.60	53 2.66 **I** 126.90	54 **Xe** 131.30
55 0.79 **Cs** 132.91	56 0.89 **Ba** 137.34	57 **La** 138.91	72 **Hf** 178.49	73 **Ta** 180.95	74 **W** 183.85	75 **Re** 186.2	76 **Os** 190.2	77 **Ir** 192.22	78 **Pt** 195.09	79 **Au** 196.97	80 **Hg** 200.59	81 2.04 **Ti** 204.37	82 2.32 **Pb** 207.19	83 2.02 **Bi** 208.98	84 **Po** (210)	85 **At** (210)	86 **Rn** (222)
87 **Fr** (223)	88 **Ra** 226.025	89 **Ac** 227.0	104 **Rf** (261)	105 **Db** (262)	106 **Sg** (263)	107 **Bh**	108 **Hs**	109 **Mt**	110 **Uun**	111 **Uuu**	112 **Unb**						

Group: 3 – 12: d transition elements

58 **Ce** 140.12	59 **Pr** 140.91	60 **Nd** 144.24	61 **Pm** (147)	62 **Sm** 150.35	63 **Eu** 151.96	64 **Gd** 157.25	65 **Tb** 158.92	66 **Dy** 162.50	67 **Ho** 164.93	68 **Er** 167.26	69 **Tm** 168.93	70 **Yb** 173.04	71 **Lu** 174.97
90 **Th** 232.04	91 **Pa** (231)	92 **U** 238.03	93 **Np** (237)	94 **Pu** (242)	95 **Am** (243)	96 **Cm** (247)	97 **Bk** (247)	98 **Cf** (249)	99 **Es** (254)	100 **Fm** (253)	101 **Md** (253)	102 **No** (256)	103 **Lw** (260)

Index